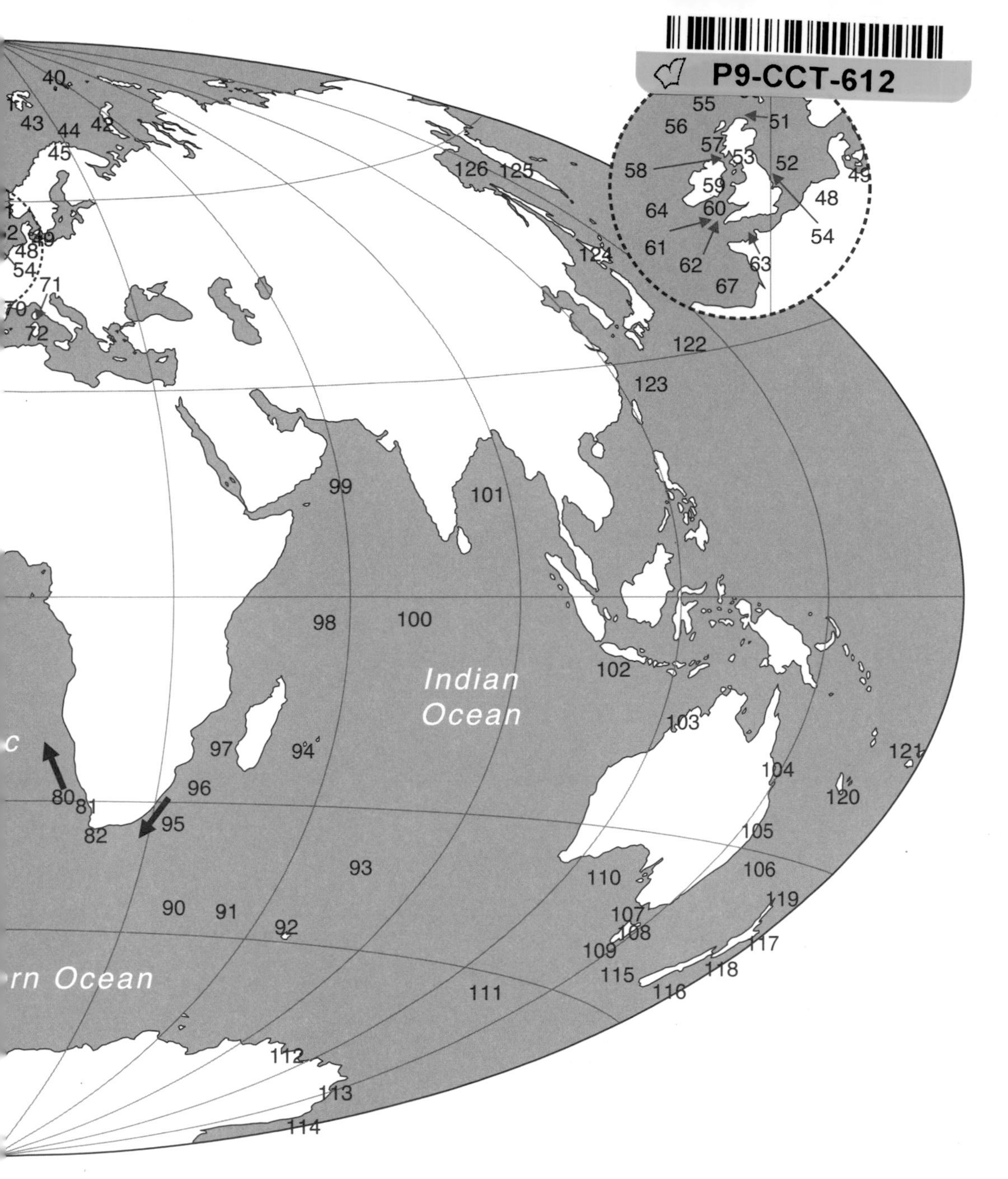


| | | | | | | | |
|---|---|---|---|---|---|---|---|
| Iles Crozet | 91 | Nasaruvaalik | 21 | Pribilof Islands | 2 | South Georgia | 87 |
| Iles Kerguelen | 92 | New Caledonia | 120 | Rangatira Island, Chathams | 117 | St Helena | 79 |
| Irish Sea Front | 59 | Newfoundland | 30 | Réunion | 94 | St Kilda | 56 |
| Isle of May | 53 | Newfoundland Basin | 31 | Ross Sea | 114 | Strait of Gibraltar | 69 |
| Kaikoura coast | 118 | North Atlantic Ridge | 65 | Rouzic, Brittany | 63 | Svalbard | 41 |
| Kamchatka | 125 | North Cape, Norway | 45 | Rum | 57 | Taiaroa Head | 116 |
| Kauai | 8 | North Carolina | 35 | Salvage Islands | 75 | Tasman Sea | 106 |
| Kimberley Region | 103 | North Rona | 55 | Sardinia | 72 | Tasmania | 108 |
| Labrador | 29 | North Sea | 52 | Scilly Isles | 62 | Terre Adélie | 113 |
| Little Barrier Island | 119 | Northumberland | 54 | Sea of Okhotsk | 126 | Torishima (Izu Islands) | 122 |
| Lundy Island | 61 | Nova Scotia | 33 | Senkaku Islands | 123 | Tristan da Cunha | 83 |
| Macquarie Island | 111 | Novaya Zemlya | 42 | Seychelles | 98 | Vermont | 34 |
| Madeira | 74 | Nysted wind farm | 49 | Shetland Islands | 50 | Washington state | 6 |
| Marion Island | 90 | Orkney Islands | 51 | Shiant Isles | 57 | Weddell Sea | 89 |
| Marseille | 70 | Palmyra Atoll | 18 | Skellig Michael | 64 | Westman Islands | 47 |
| Mewstone | 109 | Patagonian Shelf | 85 | Skokholm | 60 | Wilhelmshaven | 48 |
| Midway Island, Hawaiian chain | 7 | Pedra Branca | 109 | Skomer | 60 | Wilkes Land, East Antarctica | 112 |
| | | Pitcairn Islands | 20 | | | | |

# FAR FROM LAND

# FAR FROM LAND

*The Mysterious Lives of Seabirds*

**MICHAEL BROOKE**

With illustrations by Bruce Pearson

PRINCETON UNIVERSITY PRESS
Princeton and Oxford

Published by Princeton University Press,
41 William Street, Princeton, New Jersey 08540
In the United Kingdom: Princeton University Press,
6 Oxford Street, Woodstock, Oxfordshire OX20 1TR

press.princeton.edu

Jacket illustration by Bruce Pearson
'The Arctic Tern's Prayer' by Mary Anne Clark quoted with the author's permission, and previously published by the Poetry Society and as a Challenge winner by Cape Farewell.



ISBN 978-0-691-17418-1
Library of Congress Control Number: 2017958318

British Library Cataloging-in-Publication Data is available

This book has been composed in Cormorant Garamond and Helvetica Neue LT Std

Printed on acid-free paper. ∞

Printed in the United States of America

1 3 5 7 9 10 8 6 4 2

*Dedicated to any and everyone who loves the sea*

# CONTENTS

# A PERSONAL PRELUDE

It is a happy fact that people following many different callings will aver that they have the best job in the world. A high-altitude mountaineer surely realises how intensely precious is life, his life which hangs by a centimetre thickness of rope as the sun rises over eastern peaks. An art historian can visit the National Galleries in London or Washington and thrill equally at how the fierce, even angry daubing of bright paint yields a Van Gogh masterpiece. But let me throw into the debate another, perhaps unexpected occupation, the seabird biologist. What could be more exciting than visiting stupendously dramatic far-flung islands, encountering exquisite creatures, and trying to answer questions about their daily lives.

In brief support of my case, I proffer some personal highlights from the past 40 years. Friends and colleagues in the same trade could easily offer matching stories, occasionally blemished by tragedy stemming from the ever-present menaces of immense cliffs and unforgiving waters.

My first serious encounter with the milling throngs of a seabird colony occurred in that interval between school and university. For six weeks I served as seabird assistant on Fair Isle in the far north of the United Kingdom. There were birds to be ringed. First catch your bird. As we entered a stinking cave at the base of Fair Isle's immense cliffs, a European Shag left its nest in the inner darkness and flew towards daylight. When it passed my companion, he extended a strong arm and caught the Shag by its long extended neck. History does not record whether captor or captive were the more surprised. But the Shag was none the worse for its irregular interception, and I was hooked.

A few years later an undergraduate expedition took me to the Shiant Isles in the Minch, the channel between the Inner and Outer Hebrides. So rich are these seas that catching Pollack for supper is a quicker endeavour than visiting the fishmonger at home to buy less-fresh Pollack fillets. Owned privately by a family who generously see themselves as custodians of the islands and place few restrictions on access, the Isles offer a world class spectacle. Razorbills and Common Guillemots galore nest amid tumbled basalt boulders the size of a decent room. Atlantic Puffins circle like swarming bees above the boulders, sometimes clockwise, sometimes anticlockwise. There I learnt the truth about puffins. They may draw tourists by their clownish appearance and ability to grasp tens of sand eels in a single beakful. But they are horrible to handle. The beak is strong and sharp, as are the claws. It is all but impossible to hold them in a way that leaves one's hand safe from biting beak or scratching claws.

Having censussed the puffins of the Shiants, probably Britain's second largest colony, and completed a doctoral thesis on the biology of another burrow-nesting seabird, the Manx Shearwater, I was offered the opportunity to go south. Who could resist six months on Marion Island in the Southern Ocean? On rainy days, a huddle of chocolate-coloured King Penguin chicks looked as unreservedly miserable as a Highland bull presenting its rear to the driving squalls of a Scottish winter. When the showers had passed, shafts of sun slanted onto a gathering of tens of thousands of noisy adult penguins, picking out their black faces, orange ear patches and saffron-yellow upper chests. How I enjoyed the miraculous merging of strident cacophony and vivid colours against snowy peaks.

A few years later I scaled up 1,000 metres to reach the fern forest topping Isla Alejandro Selkirk in the Juan Fernández archipelago west of Chile. My mission: to census the Juan Fernández Petrels, whose entire world population, nourished by flying fish and squid, nests on this one island. For the better part of two weeks I clambered among the 4-metre ferns, counting burrows by day. By night I thrilled as a million pairs, give or take, visited the colony and, flying boldly into the swirling mists, crashed through the fern canopy to land near their burrows. It was towards the end of their laying period and some birds had failed to lay in burrows, as per nature's blueprint. Instead, caught short, they laid on the ground. The egg was doomed and so I gathered enough eggs to thrive on a daily omelette, cooked on a small smoky fire just outside the tent. But the ground under the ferns was peaty, and eventually started to smoulder. If my activities set the colony alight, that would be catastrophic. Nearby, water was scarce. The only solution was to stand up . . .

Ten years later I headed south again out of Cape Town towards the Norwegian sector of Antarctica. After navigating the big grey seas of the Southern Ocean, the Norwegian research vessel, R/V Polar Queen, thumped through the Antarctic pack ice for several days. I never quite became accustomed to the clattering noise of ice sliding past the vessel's hull so it was with relief that we reached the permanent ice - and could go no further. There we boarded a helicopter and flew inland for 50 minutes over white nothing. Abruptly, stony snowless mountains reared out of the ice. We landed on the margins of the world's largest known colony of Antarctic Petrels, 200,000 pairs apparently happy to nest on the snow-free slopes, albeit at least an hour's flying from the sea. Did this distant site protect the birds from the rapacious predatory South Polar Skuas? To answer that question was the objective of our study.

These are some highlights from a lifetime as a seabird biologist. My work has mostly entailed studying the birds ashore, and remaining frustratingly ignorant of the birds' activities at sea. Only over the last 20 or so years have modern electronic devices begun to reveal the details of those activities. The revelations are astounding. They have enhanced the wonder of seabirds. This book aims to share that wonder. It feels as if seabird enthusiasts have suddenly found the long-sought key to the door that allows an escape from dark ignorance to a new vista of wondrous knowledge.

# FAR FROM LAND

**CHAPTER 1**

# Introduction to the World's Seabirds

## Past Knowledge and New Revelations

I remember, at a recent international conference, a seasoned researcher receiving a medal celebrating her distinguished career in seabird research. Her cheeks had the sheen of a farmer's, well-polished and apple-red from exposure to the Scottish wind. She recounted how she had sat atop the islet of Ailsa Craig during her earlier doctoral studies and wondered where the gannets, nesting on that granite lump in the Firth of Clyde, went when they flew beyond the horizon. She had no idea, and nor then did anyone else.

But now, increasingly, they do. Modern electronics are revolutionising our knowledge of the activities of seabirds at sea. Just as mobile phones were unknown 50 years ago and the early clumsy 'bricks' clutched by Gordon Gekko in Wall Street now seem laughable compared to the latest iPhone, so it is with the electronic devices that scientists attach to seabirds. They have become smaller and more sophisticated, and opened up the watery world of seabirds to our fascinated gaze.

Seabirds can be seen in so many circumstances, all of which raise questions. Think of the father and his small son about to start eating their fish and chips on the sea wall at St Ives in Cornwall. Suddenly a Herring Gull swoops and snatches a helping of chips for its tea. The small boy is scared, the father is resigned but curious. Does that gull making its living entirely from pirating holidaymakers?[1] And what does it substitute for chips outside of the holiday season when the esplanade is empty?*

The next day the pair join a local fisherman and head offshore to catch mackerel. Catching mackerel on a handline of colourful flies may not be the most sophisticated angling, but what a thrill for a ten-year-old. The thrill is only compounded when a group of Northern Gannets surrounds the fishing smack and begins to plunge into the water. At the moment of impact, the black tips of their wings are stretched so far back as to extend beyond the tip of the tail. The birds are obviously becoming as streamlined as possible. Not only does this reduce the risk of bodily damage but it also enables them to increase the depth they reach. But do they catch the mackerel on the downward plunge or on the subsequent ascent (the latter, it turns out), and what depth do they reach? How does that depth compare to the depth a penguin attains on a dive lasting some ten times longer?

In the west of Ireland, hard-core birdwatchers barely sleep through a September night. A deep depression, the residue of a Caribbean hurricane, is passing through, rattling the windows of their hut. They will be up at dawn and quickly positioned at the cliff edge, telescopes trained on the horizon. They like nothing better than Joseph Conrad's "westerly weather . . . full of flying clouds, of great big white clouds coming thicker and thicker till they seem to stand welded into a solid canopy."[2] Their

* To share a day in the life of a Dutch seagull as it raids urban back yards, and then takes a dip in the sea, visit https://vimeopro.com/south422/animal-gps-track-animation/video/33587018.

hope is that the westerlies will have blown rare seabirds from further west in the Atlantic towards the Irish coast. These might include Great Shearwaters whose breeding home is the Tristan da Cunha group of islands of the South Atlantic. But the shearwaters passing Ireland are only a minority of the millions heading south at this season. What is the normal route of the shearwaters when they head north from their breeding grounds to spend the northern summer in the North Atlantic and then return south in September? Do they follow the same route north- and southbound, or do their travels take them on some sort of circular loop, the better to exploit prevailing winds? Do they travel continuously when migrating, or stop off for a week or more at oceanic 'oases' where the pickings are particularly good?

Forward a few months to the month of January, to the grey waters off Newfoundland where many Great Shearwaters passed by in late summer. The weather is grim, the nights long. Yet this is a part of the world chosen by many seabirds from Greenland, for example Brünnich's Guillemots, to spend the winter. To catch food, the guillemots dive many metres below the surface. Even in the middle of a winter's day, light levels and hence visibility will be poor at the depths where guillemots catch food. What allows them to succeed, as assuredly they do, and do they feed at night, when the difficulties are presumably still greater?

If guillemots face daunting dives, spare a thought for Emperor Penguins. Once a female has laid and left the male to incubate the egg through the darkness, the blizzards, the numbing -40°C chill of the Antarctic winter, she heads north to seek food in open water. But available light will be very limited, especially at depth and even more so if she dives under floating ice. Catching fish would certainly be easier if the fish (rashly) signalled their presence by flashing lights.

Further north in the Southern Ocean, the westerlies are roaring through the stormy latitudes of the forties and fifties. This is the domain of albatrosses. If there is no wind, they sit becalmed on the water. Flapping is not their forte. But let the wind blow. Let the albatrosses spread their wings and lock them open using a special skeletal mechanism. Then the birds, be they the smaller mollymawks, or the giant Wandering and Royal Albatrosses with a 3.5 m wingspan, can glide. A wind of 50 knots is no buffeting enemy; it is a source of free energy. It helps the birds to cover immense distances and to provide cheer for lonely sailors a thousand miles from land. Despite their ability to bring joy in the midst of emptiness, albatrosses have not always been treated kindly. Little heeding the fate of the Ancient Mariner, nineteenth-century emigrants bound for Australia regularly tormented and killed albatrosses as they traversed the Southern Ocean, as immortalized in Charles Baudelaire's *L'Albatros*. Sailors used the webs of the albatrosses' feet to make tobacco pouches and the wing bones to make pipes. Yet whatever the circumstances of the encounter, the seafarer surely wondered. Where do these albatrosses nest? How do they return home against the unrelenting wind if their outward journey had taken them far downwind? Or do they follow the tactic of the tea clippers and circle the globe, forever chased by the west wind?

North of the albatrosses' home of grey-green productive waters, churned by the wind, lies the blue zone of the subtropics. Look down into the limpid water from a small yacht and fancy that the water is so clear as to allow a peep into the miles-down deep. Yet the water is clear for a reason. It contains few nutrients, such as nitrates, and consequently there is little planktonic growth to cloud the water. Creatures higher up the food chain are correspondingly scarce, and so a day at sea can be overwhelmingly boring for a birdwatcher. A single petrel, the size of a small gull, arcs over the horizon, but the view is too brief to permit discrimination among several rather similar species. And that's it for another day. Even here in the midst of emptiness, the ornithologist wonders: can that lone petrel make a living in these barren waters, the blue water desert, or is it using its power of economical flight to at least seek out regions where the seas are more productive and its prey, small squid, more easily found?

* * *

Perhaps the next logical step in this tale would be to recount how far the traditional observer has taken this story. I am thinking of the seawatcher peering into the storm from a headland or the researcher, stuck unwashed on an island, who unravels the breeding habits of a seabird species with the help of binoculars, notebook and a healthy dollop of scientific intuition. This step must be postponed until I have introduced the *dramatis personae*, the world's seabirds. Among the global total of around 10,000 bird species, the seabirds are the 300–350 species that feed along the coast or out to sea, in some instances thousands of kilometres out to sea.

Luckily, the flippered penguins need little introduction. Ranging in size from the 1.2 kg Little Penguin, about the weight and shape of a magnum bottle of champagne, to the 40 kg Emperor Penguin, the 18 flightless species breed from the Galápagos Islands on the Equator in the north to Antarctica in the far south.[3] All are clad in a tight waterproof plumage that is dark above and white below, a pattern that may be helpful in camouflaging the penguins from their prey. Most colonies are on remote islands but there are exceptions: penguins breed, for example, on mainland South Africa, on New Zealand's South Island, and on Antarctica.

The largest seabird group comprises the tube-nosed birds in the order technically known as the Procellariiformes. This order, containing both highly aerial species that feed at the surface and others that are more or less adept divers, is divided into four families. One family, Diomedeidae, contains the charismatic albatrosses. Most species, 17, are found in the Southern Ocean, and there are another four confined to the North Pacific, while the final species, the Waved Albatross, mostly nests in the Galápagos and feeds off the coast of Peru. All have the long narrow wings that make for efficient gliding and the ability to cover huge distances while spending little energy.

Another worldwide tube-nosed family is the Procellariidae, comprising some 90 mostly mid-sized species. In plumage they are a motley crew; some species are all dark-brown or black, some wholly white, and others dark above and white below, maybe with a distinctive pattern on the underwing.

Within this family is a group of seven species including the fulmars and giant petrels, whose large hook-tipped bill is well able to rip open a seal carcass. These birds nest in the open at higher latitudes where burrow-nesting may not be an option – it is impossible to dig into frozen ground – and so the chick often protects itself by spitting oily vomit at would-be predators.

A
B

The petrel family named Procellariidae contains a variety of mostly mid-sized species. These include (A) shearwaters exemplified by a Scopoli's Shearwater photographed against the Mediterranean Sea carrying a geolocator device on its leg (© Maties Rebassa), (B) the extremely oceanic gadfly petrels such as Murphy's Petrel of the Pacific (© Michael Brooke), (C) the fulmars and allied species such as the Southern Fulmar (© Richard Phillips), and (D) the prions, a group where the several species look very similar. Illustrated is an Antarctic Prion (© Oliver Krüger).

Then there are the shearwaters, named because their graceful flight intermingles bursts of flapping with glides when the wing tip seems to touch and indeed may touch the water. The species breed - mostly in burrows - in temperate and tropical latitudes both north and south of the Equator.

The gadfly petrels, also primarily burrow-nesting, are another large group within the family. Their lively helter-skelter flight includes high arcs that take the bird many metres above the sea. Perhaps the high point of the arc, when the bird is on its side with wings vertical, is an opportunity to spot other petrels that have found food, or a chance to smell food from afar. That food is often squid.

A final group in the family are the prions, quite small and dull grey with flattened bills containing combs that serve to sieve plankton, especially crustacea, from the surface waters. Prions are confined to the Southern Ocean.

Also conspicuously tube-nosed are the storm petrels, now placed in two families, the Oceanitidae of the Southern Hemisphere, and the Hydrobatidae of the Northern. All of the 25 or so species are small, weighing in at between 20 and 70 g, and often black with a stand-out white rump. In other words, the smallest species, the Least Storm Petrel, is outweighed by a skinny House Sparrow. To spot such small birds pitter-pattering on thin legs over the sea surface in the slightly sheltered troughs of a 10 m swell, while the storm flails white spume off the wave crests, is to enjoy a brief respite from seasickness.

Finally, among the tube-nosed birds, the four diving petrel species (traditionally in the Pelecanoididae) are restricted to the Southern Hemisphere. With chubby body and whirring wings, used for underwater propulsion, they are remarkably similar to their northern ecological counterparts, the smaller auks, which will be introduced shortly.

The three gannet species are familiar large white seabirds with black wing tips, 'dipped in ink'. One species dwells in the North Atlantic, another off South Africa, and the third in waters adjoining Australia and New Zealand. While they are essentially temperate in distribution, their close allies, the seven booby species, are tropical. Booby of course also means 'duffer', and boobies never appear the smartest birds on their

The bare-skinned red throat of a male Magnificent Frigatebird is inflated to attract a mate.

beach, especially when showing off their brightly-coloured feet of which they seem unreasonably proud. Gannets and boobies commonly plunge from a height into the sea to feed.

The three tropicbird species are all birds which plunge to catch their prey. They are exclusively (and predictably) tropical and have mainly white plumage, adorned by a pair of spectacularly-long central tail feathers, white in two species and red in the third.

Also tropical are the five frigatebird species which are predominantly black. By way of sexual ornamentation, mature males have red throat pouches that can be inflated to attract females. Since their legs are tiny, frigatebirds are virtually unable to walk, but the reduced undercarriage and the large angular wings mean that their wing loading, the weight of bird supported by each square centimetre of wing surface, is the lowest of all birds. This gives them extreme agility, well displayed when they are chasing other seabirds, forcing them to regurgitate, and then catching the vomited spoils in mid-air before they splat into the sea.

There are about 35 species of cormorant or shag. Because various different species with vernacular names of cormorant and shag are placed in the same scientific genus, it is fair to say that there is no defining difference between the two. Perhaps this is confirmed by the first two lines of Christopher Isherwood's ditty celebrating "The common cormorant (or shag) lays eggs inside a paper bag."[4] With a worldwide distribution, these are familiar dark birds, the size of a small goose. Because of poor waterproofing, they often hang their wings out to dry after a period of swimming which involves dives from the surface to catch food underwater. While most of that food is marine, a handful of cormorant species uses freshwater habitats.

Various pelican species may visit the sea, but only one, the Brown Pelican, is wholly marine. It is a resident of the Atlantic and Pacific coasts of the Americas roughly from the Canadian border south to Venezuela and Peru.

The roughly 100 species of gull and tern are familiar. No wonder. They are extremely widespread, breeding on remote islands in all oceans, mainland coasts and well inland. They are found from the high Arctic (Ivory Gull) to the milder margins of the Antarctic continent (Antarctic Tern). Characteristically long-winged, they tend to be shaded grey above and white below. When at sea, they catch prey at the surface or by plunges that take them to no great depths. This latter habit is particularly the tactic of the terns.

The seven species of skua, all mostly brown in plumage, are allied to the gulls and indeed are gull-like in size and shape. Some species are essentially terrestrial during the breeding season. For example, the Long-tailed Skua (= jaeger) then eats lemmings on the Arctic tundra, and some South Polar Skuas are specialist predators at the colonies of Antarctic seabirds. When not breeding, skuas largely remain at sea. How much of their living is made by piracy of other birds and how much by independent feeding remains uncertain.

The auks are a family of seabirds confined to the Northern Hemisphere, with a stronghold in the North Pacific. The 24 extant species are specialist divers, as was the extinct flightless Great Auk, and they can be thought of as the ecological equivalents of the southern hemisphere penguins. Both groups use their wings (or flippers in the case of penguins) for underwater propulsion when hunting prey, often at remarkable depths (see Chapter 9). However, crucially, the living auk species can all fly as they are distinctly smaller than the penguins, ranging in size from around 85 g (Least Auklet) to 1 kg (Brünnich's Guillemot*).

That almost closes the curtain on the *dramatis personae*. Nevertheless there are other birds that routinely use the sea. Think of them as the courtiers and countryfolk of a Shakespearean cast. They adorn the stage but contribute little to the narrative. All the divers (= loons) and some grebes are marine outside the breeding season. This pattern is followed by a number of ducks, whilst the eider ducks are marine throughout the year. Finally two of the three phalarope species are essentially marine when not breeding and can be seen bobbing cork-like in such places as the Arabian Sea and among the Galápagos Islands. For reasons of convention as much as logic, these species are generally not considered seabirds, and they will make only the briefest appearances in the chapters that follow.

* * *

The introductory pages raised questions about the activities of seabirds as they go about their daily and nightly business in their watery realm. My hope in this book is to describe how far modern gadgetry, much of it electronic, has enabled enthusiastic researchers to answer these questions. Before embarking on this exciting tale of revelation, it is worth recounting quite briefly the sort of information that has been the mainstay of seabird research in the past.

Centuries of observation on land and at sea have yielded a fair picture of how many species of seabird there are. Nonetheless surprises still occur when supposedly extinct species are found to persist and wholly new species are discovered. As recently as 2008 Monteiro's Storm Petrel was described from the Azores, to be followed in 2011 by Bryan's Shearwater from Midway Island in the Hawaiian chain. But these two cases are complicated by the fact that the birds were known in earlier years. Only the availability of new evidence, on timing of breeding, DNA, and fine-grained plumage features has allowed the description of new full species. The most recent, more dramatic, announcement happened in

* Known as Thick-billed Murre in North America

2011 when Peter Harrison, a doyen of seabird identification, announced the discovery of a brand-new species, the Pincoya Storm Petrel, that flits over the fjords of southern Chile.[5] It had escaped the notice of Charles Darwin who had sailed those waters aboard the *Beagle* almost 200 years earlier.

Once the who's who of seabirds has been established, it begins to become possible to establish broad migration patterns. Consider, for example, Great Shearwaters, an 800-gram species whose stronghold is the Tristan da Cunha group of islands in the South Atlantic. There they are harvested by Tristan Islanders, and I can vouch for the superlative chips made from potatoes nurtured in the islanders' potato patches and fried in shearwater fat. It has long been known that the Great Shearwaters appear in force off Newfoundland in the northern summer, and it is obvious they migrate between North and South Atlantic. That leaves unanswered a multitude of questions about the speed and precise route of the journey.

While such simple observations have been a source of knowledge about seabird migrations, surprising gaps have persisted. Atlantic Puffins, much photographed with a beakful of fish at their colonies, all but disappear in the winter despite being one of the most numerous seabirds of the North Atlantic. They must be all at sea somewhere. Conversely Hornby's Storm Petrel, a pale grey sprite found 40–300 km off the coasts of Peru and Chile, is a common bird of the cool waters of the Humboldt Current. American ornithologist Frank Chapman, quoted by Robert Cushman Murphy,[6] the long-serving Curator of Birds at the American Museum of Natural History, describes the petrels on the wing as "the most erratic flier[s] I have ever seen . . . like a bat, swift and nighthawk in one." Young Hornby's Storm Petrels on their first journey from the nest to the sea are regularly attracted to the lights of Chile's northern desert cities, implying the colonies cannot be far away. Yet no-one has ever found a colony of this species, probably the world's commonest seabird whose breeding places are wholly unknown.* Less surprisingly,

* In April 2017, while this book was in press, an active colony of Hornby's Storm Petrel was finally discovered in the Atacama Desert 70 km from the coast. See http://www.redobservadores.cl/equipo-de-la-roc-encuentra-el-primer-sitio-de-nidificacion-de-la-golondrina-de-mar-de-collar (accessed 14 June 1017).

the colonies of several other very rare species have only recently been discovered or remain unknown. A colony of the Chinese Crested Tern was first discovered in 2000, but the species' world population is tiny, perhaps fewer than 100 individuals. MacGillivray's Petrel of the South Pacific may be equally rare. It probably nests somewhere among the Fijian Islands but no-one knows where.

Once the rough picture of the cast of seabird species and their global distribution has been painted, studies at colonies – and nearly all species nest in colonies – can start to flesh out details of the birds' breeding habits. Most obviously, they reveal when the species breeds and how long it takes to incubate eggs and raise young. That said, it is remarkable how little was known just a human lifetime ago. For example, Ronald Lockley, a pioneer of seabird research, studied Manx Shearwaters, a 500-gram species dear to my own heart, on the Welsh island of Skokholm before World War II. He had no idea how long they incubated their eggs. He continues the story[7] "On the fiftieth day our shearwater had beaten all records for incubation that I, at least, had heard of. The white stork takes 30 days . . . and the tame swan 38 days, usually less, to incubate its eggs . . . and even the vulture takes only 48 days. . . .

Ada [the female] was on the egg on the fifty-first day. I had determined to test the egg by gently shaking it. . . . There was no need for any test with Ada's egg that morning, however. To my delight it was pipped. Next morning. . . . Ada was brooding the chick. . . . The egg had taken 52 days to hatch and so had made a record for length of incubation of a fertile egg laid and brooded by a wild bird."

Since then, the incubation periods of many species have been determined. They range from just under three weeks among the smaller terns to about 10 weeks in the great albatrosses. Rearing the chick to fledging can be correspondingly protracted, about 10 months in the largest albatrosses. These spans of parental care are decidedly longer than in most landbirds.

More intriguing information emerges when birds are ringed (or banded) and given an individual identity. It transpires that the great majority of seabirds remain faithful to the same partner year after year, the pair bond being broken by death or the occasional divorce.

Ringing is also a powerful tool for assessing survival from one year to the next. Despite often spending most of their lives in seas apparently so hostile, seabirds are actually rather long-lived, and researchers, the people who love nothing more than to smell the guano, expect their marked birds to return to the colony year after year, and become old friends. Ninety-seven percent of adult Wandering Albatrosses survived from one year to the next in the days before longlining added a fearful extra threat to these maestros of gliding. In contrast, when just three-quarters of adults survive the year – the case for Common Diving Petrels – the survival figure seems low to an experienced seabird ornithologist, despite being high in comparison to the survival of a garden bird in a temperate country.

If seabird survival is high, it might be anticipated that the number of young produced by breeding birds would be low. Were that not the case, the oceans would be awash with birds. And, indeed, low output is the order of the day. The great ornithologist David Lack pointed out how species that range furthest over the oceans tend to lay a single egg.[8] The presumption is that bringing enough food for a single youngster is hard enough work for these parents, which of course helps explain the very long fledging period of albatrosses. It is only species feeding inshore and fairly close to their colonies that lay larger clutches, for example two or three eggs are laid by certain gulls and terns, and up to five by some cormorants.

Another piece of this introductory jigsaw is the observation that the most oceanic, wide-ranging species laying a single egg tend to nest in colonies that are far from one another, and sometimes huge. Sooty Tern colonies can exceed one million pairs. On the other hand, the species feeding closer to shore nest in smaller colonies that may be sprinkled along a coastline at no great distance from each other.

Once a seabird has a lifestyle that promises many years on the wing, natural selection will set to work. In particular natural selection will favour individuals which do not imperil their own long-term chances of survival by recklessly over-investing in any single year's offspring. It would simply be counter-productive to die in the defence of one chick, and fail to survive to rear many chicks in future years. This line of argument helps explain the small clutches of seabirds. It also bears on the age at which seabirds start to breed. While some species begin breeding at two years old, between four and six is commonplace. Life in the windy lane of the Light-mantled Sooty Albatross is so leisurely that, on average, the birds do not start breeding until the age of 12. This general pattern is part and parcel of a life history involving high adult survival. Birds do not start breeding until several years have passed. Those years may be needed for them to acquire the maritime feeding skills necessary to take on the extra burden of feeding a chick. Or they may need to acquire knowhow to minimize the additional hazards encountered when visiting land. A storm petrel dismembered and crushed in an owl pellet, a shearwater with a broken neck after a night-time impact with a rock face at the colony: these are nature's failures.

Another aspect of seabird biology much illuminated by colony studies is diet and feeding habits. Sometimes the findings can be guessed in advance. Food ferried from sea to colony in the bill is likely to have been caught close at hand since this mode of transport is aerodynamically inefficient. When a tern feeds its chick a sand eel that is still glistening salty wet in the bird's bill, it has obviously been caught nearby. Food brought from further afield is likely to be regurgitated to the chick by the parent, and the inquisitive ornithologist can persuade the unlucky bird to regurgitate its hard-won catch into a collecting vessel, there to be sifted and identified. Sometimes such observations lead to surprising conclusions. French researchers Henri Weimerskirch and Yves Cherel studied Short-tailed Shearwaters breeding on Tasmanian islands.[9] Some of the food, krill and fish characteristic of colder seas, brought back to chicks after the adults' longer trips, indicated that the birds were travelling at least 1,000 km south of Tasmania into Antarctic waters to forage. When New Zealand ornithologist Mike Imber noticed that the food brought by Grey-faced Petrels to their chicks was substantially made up of squid species that migrate from the depths towards the surface at night and there emit light, he wondered whether the petrels actually fed at night, perhaps targeting the glowing molluscs.[10]

Whilst the type of food brought by seabirds to their colonies certainly allows inferences about where that food was caught, so, sometimes, does the length of the foraging absence. A tern that returns to the colony after an hour's absence gripping a fresh sand eel has not gone far. At the other extreme, some petrels and albatrosses sit on their egg for 20 days whilst the mate feeds at sea before reappearing to resume incubation duties. There evidently has been time enough for the mate to cover immense distances – but did he or she go north, south, east or west? Did the outward and return journeys follow the same route, or was the overall track a loop? Can the absence be split into obvious and distinct travelling and feeding phases? Only the recent arrival of tracking devices has begun to provide answers to such questions.

Seabird biologists love to count seabirds, to the extent that we now have a tolerably accurate estimate of the number of breeding pairs of most species. Make some assumptions about how many immature birds there are in the queue to join the colony, and it is possible to estimate the number of each species on the wing. Add them together and there may be some 700 million seabirds on earth, about one-tenth the number of people. Especially numerous are the diving species of higher latitudes: the penguins, shearwaters and auks. Knowing the food requirements of birds of various sizes, it is possible to calculate the aggregate amount of food they extract from the sea in a year. The total of at least 70 million tonnes is remarkably similar to the amount, some 80 million tonnes, the fishing industry has brought ashore each year since 2000.[11]

This sketch of the sort of knowledge seabird biologists have accrued from land-based studies might, one would think, be expanded by observations at sea. That is true to a degree. When North Sea oil production was getting underway in the 1970s, numerous surveys were undertaken to assess which parts of the basin were preferentially used by seabirds that might fall foul of oil spills. Although such information may help conservation planning, it reveals nothing about the origins of the birds seen. In the same era, Pierre Jouventin and colleagues travelled south from the French Département of La Réunion, in the tropical Indian Ocean. Heading south towards Antarctica, they showed how certain albatrosses, such as Wandering and Indian Yellow-nosed, and the Great-winged Petrel, were seen most frequently near two zones where sea temperature altered abruptly, the Subtropical Convergence and the Antarctic Polar Front.[12] This implies that the birds were seeking out these zones of water-mixing and therefore enhanced marine productivity for feeding (see Chapter 8). It tells us nothing about the colonies from which the birds hailed, or whether they were breeders or non-breeders. Again it is recourse to tracking gadgetry that leads towards answers to these finer-grained questions.

I remember crossing the North Sea from Newcastle to Oslo in January 1968 aboard a smart passenger ferry. For a large fraction of the limited daylight I was wedged in a secure nook astern. The air was chill, the ship's wake serpentine, twisted by the lumpy waves. And the Northern Fulmars enchanted me, gliding this way and that with no apparent care in the world. Of course, they did have a care; the imperative need to find food. And that leads to the persistent worry about such surveys as those from the North Sea, and from the Indian Ocean mentioned in the last paragraph. They recorded the presence of auks and fulmars, and albatrosses and petrels respectively, but it can be quite rare for observers to see the birds actually feeding. Is this because the birds manage to catch enough food to last, say, a couple of days during infrequent bouts of gorging, or is it because much feeding happens at night when they cannot be seen? Devices that tell us when birds actually ingest food have the potential to provide an answer.

Without question, birds follow ships in the hope of grabbing food. That might be galley waste from a yacht but of course fishing vessels are potentially the richest source of food. Sometimes this is offal thrown overboard after the fishers have gutted the catch, or it could be discards, fish thrown away because they are of no commercial value. Sometimes birds target fish leaking out of a trawl as it is retrieved, and put themselves in danger from the taut trawl wires. Even more perilously birds are attracted by the baited hooks that are accessible while a longline of several kilometres is being set. As the line streams astern, there is a short time window when each baited hook can be grabbed by a bird before that hook goes too deep to be reached. If a bird grabs the bait, it may be lucky and win a meal. It may be unlucky. It gets hooked, is dragged underwater, and drowns. For some species, such as the Northern Fulmar, food sourced from fishing vessels has been hugely important and a major driver of twentieth century population growth. Just what proportion of the diet of a typical individual fulmar is derived from this source is less clear. For other species, for example the Southern Ocean gadfly petrels which barely interact with these vessels, it is of no importance. The overall picture then is that seaborne observations may be giving a biased picture of the feeding habits of some seabird species, and no picture at all of other species. Can modern technology help steer us away from such biases?

* * *

Until the middle of the twentieth century, it was possible to watch seabirds on land and at sea, and to study them in more detail at colonies. The latter activity included, *inter alia*, putting metal rings on the birds' legs, a well-established means of studying bird migration. In the case of seabirds, those bearing rings could be recaptured at the colony or, possibly, found dead on some more or less distant shore. How informative is that washed-up carcass? Had the bird strayed from its normal route, encountered barren seas, and died? Had it drifted as a corpse some hundreds of miles from the position of death? While doubt clouds the picture painted by dead ringed birds, modern tracking devices yield much higher quality information about birds' whereabouts.

Attaching transmitting VHF radios to animals has occupied biologists since the late 1950s. It is a powerful technique for relocating, say, a troop of chimpanzees that assuredly will not have travelled far since their last known position. It is less useful for seabirds which travel far greater distances, taking them beyond the line of sight of any scientist deploying a receiving aerial on some windy clifftop. Couple this problem with the fact that a seabird will often dip into the trough below the wave crests or, even worse, submerge underwater, and the upshot is that VHF radio-telemetry has not transformed seabird research.

Those disparaging words notwithstanding, radio-telemetry has had its moments. In 2003, the ornithological world was amazed when the New Zealand Storm-petrel, thought extinct for over a century, was rediscovered at sea off New Zealand's North Island. That led immediately to the question of the whereabouts of its colonies, and the tricky task of discovering those colonies. The problem was solved when it proved possible to attract the birds close to a 3.5 m inflatable with chum, the ornithologists' term for a smelly sludge of fish bits. Once in range, the storm-petrels were captured by a small net fired over them. Fitted with

Believed extinct for over a century, the New Zealand Storm-petrel was re-discovered in 2003. Subsequently, radio-tracked birds led scientists to a colony near Auckland.

a transmitter weighing two-thirds of a gram, the released birds then led the searchers in 2013 to nesting burrows in the rainforests of Little Barrier Island, a mere 50 km from Auckland, New Zealand's largest city.[13]

Other techniques may bear future fruit in the search for breeding sites of other species whose colonies remain unknown. For example thermal imaging and radar have helped pinpoint nesting areas of the Black-capped Petrel in the mountains of Hispaniola, and given hope that the species, long thought extinguished from the Caribbean island of Dominica, still breeds there. Drones carrying a thermal-imaging camera may contribute to identifying the whereabouts of Marbled Murrelets nesting under the dense canopy of the old-growth forest running along the western seaboard of North America.

The overall impacts of VHF radio-telemetry and radar have been slight compared to what has been learnt from satellite telemetry! The first successful deployment of satellite transmitters (on any bird) was achieved by Pierre Jouventin and Henri Weimerskirch in 1989.[14] They attached 180 g devices to male Wandering Albatrosses breeding on the

French Iles Crozet in the heart of the Roaring Forties. Five males were followed as they made off-duty journeys between 3,700 and 15,200 km while their mates incubated the single egg and awaited the return of the wandering males. These values represent minimum journey distances since the satellite passed overhead to collect positional information every 90–100 minutes, and it was assumed, conservatively and probably incorrectly, that the bird had followed a straight line between the two points.

Since these pioneer studies, devices reporting to satellites, known in the jargon as PTTs (platform transmitter terminals), have been deployed on countless species, as we shall see. Meanwhile the weight of devices has fallen dramatically, and continues to fall. Devices weighing about 5 g are readily available today. At a pinch such devices could be deployed on a 100 g bird, although the 'industry' standard is for the load not to exceed three percent of the bird's weight, especially as it is scientifically pointless and ethically reprehensible to obtain data from a bird behaving abnormally. Since a major component of the overall weight of a PTT is the battery, this has been minimized in the smallest modern devices by including a small solar panel that regularly trickles current to the reduced tiny battery.*

Because of the technicalities of how the satellite system computes position, accuracy of PTTs is around 500 m. Impressive and good enough for many seabird studies, but poor compared to GPS accuracy. GPS (global positioning system) entered the public domain in the late 1980s. Come the summer of 1993, the US launched the 24th Navstar satellite into orbit. That completed the modern GPS constellation of satellites, 21 of which were active at any one time, leaving three more as spares. Today's GPS network has around 30 active satellites in the GPS constellation, delivering an accuracy of comfortably below a metre to the building industry, navigators and many more users. This accuracy is a boon to seabird researchers, especially as the smallest devices weigh in at around 1 g. At present this weight only allows a limited number of position fixes.

* This development has allowed the deployment of 1.6 g devices onto a number of Spoon-billed Sandpipers, enchanting and fearfully endangered tiny waders that migrate between north-eastern Siberia and south-east Asia.

Increase the weight to around 3 g and add a solar panel trickle charger to power the tiny battery to deliver more fixes.

Superlative accuracy has come at a price. The scientist has faced the need to recapture the subject in order to download the data stored on the GPS tag. That can be problematic if the seabird has learnt that scientists bearing nets or hooks are best avoided. Even this constraint is starting to dissipate. Some GPS devices will 'talk' to satellites. Others will transmit the stored data to base stations set out in the colony to which the birds will assuredly return. Those base stations then send the data to the ornithologists via the mobile phone network.

The accuracy available is astonishing. I am especially fond of an animated online plot of the day's travels of a GPS-tracked Dutch gull, as seen from the air. After leaving its roost on a coastal sandspit, the gull heads a short distance inland, visiting a succession of urban backyards. The day's feasting done, the bird washes away the accumulated grime with a dip in the sea before returning to roost on the very same sandspit.[15]

Finally positional information can be gathered from geolocators (also known as Global Location Sensing [GLS] trackers or geologgers), light-sensitive devices which, attached to the bird, record the time of local sunrise and sunset. This allows determination of day length and local midday, as a function of the day of the year, which in turn yield the bird's latitude and longitude, respectively.[16]

The very earliest geolocators were attached to Northern Elephant Seals on California beaches in the late 1980s, and revealed the seals headed to the North Pacific when not breeding.[17] Ten years later, device size was down to 20 g, and the flood of information from albatrosses was underway. The most modern geolocators weigh well under 1 g and can be attached to small 20 g passerine birds, giving unimagined insights into their migratory routes.

Geolocators have two snags. Latitude information is poor around the equinoxes when daylength everywhere in the world is around 12 hours, and accuracy may be no better than a few hundred kilometres. Notwithstanding these drawbacks, the relative cheapness of geolocators and their ability to run for two years or more until the subject bird is re-captured for data retrieval mean that they have been wonderfully informative. In the desk drawer beside me, as I write, are half a dozen retrieved geolocators. They have accompanied me by aeroplane from Cambridge via Los Angeles airport to the South Pacific and back. They have been deployed for two years on six Murphy's Petrels, each of which has flown from the Pitcairn Islands to the North Pacific and back, not just once but twice - and that is not to mention the birds' excursions from the nesting colony of thousands of kilometres.

Having established the whereabouts of a seabird, the next obvious question is: What is it doing at sea? The first phase in any answer might be to establish whether it is flying or swimming. Enter immersion loggers. Obviously a flightless species, such as a penguin, will have wet feet for as long it remains at sea. The picture for volant species is more complex, as we shall see in later chapters. There may be differences in the proportion of time spent swimming on the water by night and by day, and there may be differences according to season. Many smaller petrels are mostly on the wing whilst at sea during the breeding season but spend over half their time on the water when not breeding. The route towards documenting such behaviours involves immersion loggers. Commonly attached to the legs of birds, these loggers sprout two small electrodes. The impedance between those electrodes diminishes when they are in water, and the associated recorder registers the time of transitions from one state (wet) to the other (dry), and vice versa. As so often, the pioneer devices, deployed on the much-studied Wandering Albatrosses of Bird Island, South Georgia, were chunky at 24 g. Today, such immersion loggers are routinely incorporated within the GLS devices fixed to seabirds, the whole package weighing less than 5 g.

If the species is bobbing on the sea, it might well dive for food. How deep does it dive? Early in the quest for answers capillary tubes were attached to birds. Because the capillary is sealed at one end, the air within becomes compressed when a bird dives and water under pressure enters from the other end. The deeper the dive, the further up the capillary the water moves. This movement was recorded by an indicator powder (e.g. icing sugar, or water soluble dye) dusted onto the inside of the capillary that changes as it gets wet. Thus, when the device is retrieved from the bird, the capillary gives an indication of the maximum depth reached by the bird and the device during the period of attachment. Since the device is not providing a continuous read-out, the longer it is deployed, the greater the maximum depth is likely to be. Crude as this technology was, it yielded surprising answers. Who would have bet on a Short-tailed Shearwater reaching 70 m?

To document how much time a bird spent at various depths en route to the crude maximum, the next step was the development of devices that recorded, either via light- or radiation-sensitive film, the amount of time the air/water boundary was at different positions within the capillary. This was certainly an improvement but it was not the continuous record of depth over time for which the curious naturalist yearns. Such a record would, for example, allow questions about how the time the bird spends at the surface is affected by how deep it has just been, and by how deep it will go on its next dive.

The credit for inventing such a device goes to the Japanese researcher Yasuhiko Naito in the late 1980s.[18] A compressible bellows, responding to pressure and therefore depth underwater, was attached to a stylus inscribing an ultra-thin line on carbon-coated paper on a rotating drum. When the paper was retrieved, it showed the bird's dive profiles. These might be U-shaped if the bird has lingered at maximum depth, or V-shaped if it has descended smartly to maximum depth and returned equally smartly to the surface. This recording system is 'old-fashioned' analogue. The data documented by the latest time-depth recorders (TDRs) are recorded digitally.

This book will delve into the ecology of seabirds rather than their physiology. But physiology cannot be ignored. A penguin diving to beyond 100 m is putting its body through serious stress. Implanted devices can measure some of those stresses. For example, an implanted heart-rate monitor (which has to be retrieved surgically for its data to be downloaded[19]) can reveal how King Penguins, also equipped with a depth recorder, show remarkable fluctuations in heart rate during the course of a dive (see also Chapter 9). Not only does heart rate fluctuate with activity, it is also probably a good indication of the amount of energy being expended in whatever activity the monitored bird is performing. The bird's energy needs translate into its food requirements and hence impact on the marine ecosystem.

Since it will always be difficult to assess without bias when seabirds feed, especially whether they do so at night, indirect means come in handy. One such way is to insert a temperature sensor into the stomach. Remembering that seabirds are universally warm-blooded, and that their prey is cold-blooded and living in waters that are at least somewhat and usually a lot cooler than the bird's body temperature, the ingestion of prey will cause the bird's stomach temperature to drop. The larger the prey item, the greater and longer-lasting the drop.

Such a device was developed by Rory Wilson, then working in Germany, and tried out on captive African Penguins in South Africa. In the dry parlance of a scientific paper,[20] Wilson wrote "Four penguins were captured from a non-breeding group at Dassen Island . . . at 11:00h on 21 June 1991 and housed in a large wicker basket for 1 h before each was induced to swallow a [device]." Scientific persuasion was also needed to retrieve the devices from the birds – but the idea worked. The sensor showed a precipitous drop in temperature when a prey-sized (50 $cm^3$) shot of water was inserted into the penguin's stomach by catheter. The drop was followed by a gradual recovery in temperature as the 'prey' warmed back to body temperature. Later trials with free-living Wandering Albatrosses on the South African sub-Antarctic island of Marion confirmed the potential of the devices. Today's devices are often inserted in the oesophagus, instead of stomach, allowing more precise timing of ingestion events.

Some ten years later, another technique for registering underwater prey capture was developed. In fact the technique was pioneered during a study of Weddell Seals.[21] Having glued a reed-contact and magnet onto the hair-covered parts of the upper and lower flews, the fleshy outer lips of the seal, the researchers could record when electrical contact was broken, in other words, when the animal opened its mouth. If a depth-recorder revealed that the seal was then underwater, it was a definite possibility that it was opening its mouth to snap up food. The same approach has been extended to Leatherback Turtles, half-tonne leviathans that convert the watery pulp of jellyfish and comb jellies into reptilian flesh. Similar devices have since been attached to penguins and shags. With a magnet on one mandible and the so-called Hall sensor on the other, the voltage recorded from the sensor decreases as the distance to the magnet increases. Thus opening the beak wide leads to a bigger drop in voltage than does a small parting of the beak. Very likely the degree to which the beak is opened is related to the size of the food item ingested.

When oesophageal and Hall sensors are recording simultaneously from the same (tolerant) penguin, there is not a perfect correspondence between the two channels. For example, the bird may open its beak twice a few seconds apart to catch two different items but these are not resolved as different by the temperature sensor. However the overall correspondence is remarkably convincing, allowing the logging of when the bird eats, and roughly how much.

Now imagine a bird carrying not only the oesophageal and Hall sensors but also a time-depth recorder. It is often possible to spot small wiggles in the trace of a bird's depth. This less-than-technical term refers to small, quick changes in depth, exactly the sort of changes one would expect were the underwater bird deviating from a straight course to snap up prey. And indeed it transpires the wiggles coincide with temperature changes in the gut and with beak opening.[22] A wiggle provides another means of detecting when a bird consumes prey.

Presuming it is unrealistic to ask a seabird to keep a diary of its daily diet, the next best might be for it to carry a camera that records the rolling view in front of its beak. Every fish or shrimp eaten would make a smart exit from the field of view as it entered the bird's gullet. Devices attached by Yutaka Watanuki approached this gold standard.[23] His team attached cameras to male European Shags tending small/medium-sized chicks on the Isle of May off Scotland's east coast. Retrieved a day later, the cameras showed the shags diving in a mix of sandy and rocky habitats. Sometimes they returned to the surface where the camera took a picture of the prey, butterfish. However it seemed quite likely that smaller items, such as sand eels, were quickly swallowed underwater and missed by the camera which fired only every 15 seconds. Items are less likely to be missed if the bird is carrying a continuously-recording video camera as described in Chapter 9.

Another recent proof-of-concept study was led by Steve Votier of the University of Exeter. Travelling to the gannetry on the Welsh island of Grassholm, with assistance from a blisteringly powerful jetboat, Votier and team attached 45 g cameras, firing once a minute, to the central tail feathers and GPS loggers to the backs of parent Northern Gannets rearing chicks.[24] Of the ten gannets whose cameras yielded useful results, seven clearly interacted with fishing vessels, mostly trawlers, during their foraging trips. During these interactions, they took pictures not only of the ships, but also of fellow gannets and indeed other birds hoping for discarded fish and other fishery spoils. That said, less than half the time gannets spent in circumscribed food-searching mode was associated with vessels; the gannets were clearly capable of independent 'natural' foraging.

In the past 15 years, accelerometers have proved an increasingly powerful tool for investigating birds' behaviour, especially underwater, and for assessing how hard they have to work to achieve that behaviour.[25] An accelerometer is conceptually simple, and measures g-force – as is needed, for example, to trigger a vehicle's airbag that inflates during the severe deceleration of a collision. Early devices attached to penguins proved useful in describing their swimming habits and how much porpoising above the waves contributed to their journeys to and from feeding areas. These devices recorded information once a second, sufficient to describe body posture. The world has progressed and accelerometers can now record from all three mutually-perpendicular axes at a much higher frequency. A 30-times-a-second (30 Hz) frequency provides data on the beating frequency of a penguin's flipper, and how that alters in the course of a dive. A team at the Isle of May, Scotland, used a recording frequency greater than 50 Hz to show that the island's Shags needed to beat their wings ever faster in order to remain airborne as they progressively filled up with food during an excursion from the colony.[26]

Dogs are routinely 'chipped' with a PIT (passive integrated transponder) tag about the size of a grain of rice. Brian Smyth and Silke Nebel from the University of Western Ontario, Canada, describe the technology succinctly. "Essentially, PIT tags act as a lifetime barcode for an individual animal, analogous to a Social Security number and, provided they can be scanned, are as reliable as a fingerprint.... PIT tags are dormant until activated; they therefore do not require any internal source of power throughout their lifespan. To activate the tag, a low-frequency radio signal is emitted by a scanning device that generates a close-range electromagnetic field. The tag then sends a unique alphanumeric code back to the reader."[27] From a seabird perspective, the absence of a battery and its associated weight is a bonus. The need for the scanner to be close to the chipped bird is a drawback, one which may lessen in a densely-packed colony where the comings and goings of a cluster of individuals can be monitored by a single scanner. That approach has been successful in the Antarctic where David Ainley coordinated a study of Adélie Penguins.[28] The penguins were ushered over a weighbridge and past a scanner as they arrived at and departed from the colony, allowing the weight of food delivered to chicks by identified chipped penguins to be determined. Even more sophisticated has been the study of Common Terns led by Peter Becker.[29] The terns nest on six concrete islands in the harbour of the German coastal town of Wilhelmshaven. All chicks fledging from these islands since 1992 have been chipped. Those that survive to return in subsequent years find themselves monitored automatically by an electronic surveillance system of antennas on elevated platforms that remotely record individual attendance throughout the breeding season.

This section has not been comprehensive but it has described the crucial devices now available. They allow seabirds to be studied as never before, despite the obstacles imposed by their journeys covering huge distances across inhospitable seas. Seabird researchers investigating mid-sized and large species can now map where their study bird goes. They can combine garnering this positional information with attaching a device that signals whether it is wet or dry and simultaneously collects data on depth, sea temperature, and light levels. It would be exaggerating to claim that a seabird can be more closely monitored than a patient in intensive care.[30] But the level of understanding of how birds live out their lives away from the apparent comfort of land is growing in a truly remarkable manner.

This book describes that growing understanding. From the data, a picture of mastery emerges. Seabirds are not helpless morsels of life tossed hither and thither by wind and waves. Rather, they employ strategies that enable them to cover huge distances and detect scattered food with relative ease, and with the advantage that they are less subject to day-to-day predation than are landbirds. No wonder seabirds attain an age of 30 regularly, and 50 sometimes, milestones far beyond the reach of any everyday garden bird.

CHAPTER 2

# Taking the Plunge

## Seabirds' First Journeys

None of us remembers being born. But suppose the normal human trajectory was to spend the years up to adolescence in the confines of a deep dark cave, or perhaps in a tree house, there being fed by our parents. Eventually the day would arrive when we ventured forth. Leaving the cave, dropping out of the tree, would be such a life-changing transition that it would surely trace an indelible memory.

When the time to leave the nest arrives, many a seabird experiences a transition of similar magnitude. Think of a young guillemot that has spent 20 days amid the clamour and guano of another hundred chicks on a cliff ledge 80 metres above a strip of boulders sloping down to a grey-green northern sea that is restless and churning. Over the past few days the chick, now one-quarter the weight of an adult, has flapped its tiny wings increasingly urgently, oblivious of the fact that they are too small to generate level flight. One evening, partly prompted by its father's calls, it leaps, flaps, and glides. Ideally its slanting downward trajectory will lead beyond the boulders at the cliff base where a painful bounce and a Great Black-backed Gull await. The preferred outcome is certainly a frantic flutter directly to splash-down, allowing father and chick to meet on the water and swim away from land. Or imagine a young Wandering Albatross that has spent some 10 months at the nest. Fed by visiting parents once a week, it has grown out of all recognition from fluff ball to a full-sized albatross with a wing span exceeding 3 metres. For the past few weeks, it has opened its wings more and more often and felt the lift created by the westerlies of the Southern Ocean. At times, the lift almost overcomes gravity. The bird then relies on a hasty retraction of the wings to flop back down to *terra firma*. But eventually timid delay must give way to bold flight. And I shall always remember the one time I observed a young Wanderer take its first flight. It spread its wings, dithered, dithered again, and then jumped from the top of a cliff. Within seconds the breeze was supporting its wings and it flew firmly westward towards a fading orange sunset. But sometimes the aerial route is not an option for the fledging seabird. Travel south to the Antarctic where the ice shelf fringing the continent is cracking up under the warmth of the summer sun. Timing is all. Should the ice break up too early, the crèche of Emperor Penguin chicks may drift away on an ice floe and, lost at sea, be deprived of parental feeds. If too much fast ice persists, this too will impede the delivery of food to the youngsters. But if the break-up of ice follows a roughly normal schedule, the chicks will have shed their down and completed growing the stubby feathers of water-going penguins about the time open water begins to lap the fringes of their colony. Fledging then entails a splash off the ice into the chill sea, and an immediate launch into independent life.

Once a young seabird left its nest site, there was, in the past, rather little information about what happens next. Partly this was because, with most seabird species not breeding until they are several years old,

No turning back! A fledging Wandering Albatross takes its first flight.

it was the norm for immature birds of many species to remain entirely at sea, and out of reach of scientists, for a year or more. For sure, some ringed birds were found dead in those early months and years, providing significant pointers to their direction and speed of travel. But this information was quite skimpy. What has recently been discovered about the very first journeys made by young seabirds as they embark upon a life on the ocean waves?

One obvious distinction is between species that quit the colony by swimming and those that fly away. At times, this is blindingly obvious; penguins only swim. Yet this does not preclude impressive journeys. The first major journey of young Emperor Penguins satellite-tracked over two months from the Ross Sea region of Antarctica by Gerald Kooyman and Paul Ponganis took them almost 20 degrees northward, equivalent to a swim from Miami to Boston.[1] Good swimmers they may be, but the young penguins still lack the diving abilities of older birds, at least partly because the amount of myoglobin, the critical oxygen-carrying protein in their muscles, remains well below adult levels.[2]

Equally remarkable were journeys undertaken by the Emperors' smaller relatives, King Penguins.[3] Eighteen fledglings of this sub-Antarctic species were tracked from the Falkland Islands and from South Georgia. Most birds initially made for the Antarctic Polar Front, a zone of high marine productivity (see Chapter 8) midway between the Falklands and South Georgia. There is a hint here that the birds 'knew' where to go. Then, after a period spent in this zone, seven of the eight birds that were still able to be tracked four months after fledging passed through the Drake Passage into the Pacific Ocean, while the eighth turned east towards the southern Indian Ocean. With average daily distances travelled of 45 km, equalling an onshore marathon, no wonder the penguins travelled up to 4,500 km from the colony. Pride of place went to a penguin named 'Youngster' by the tracking team. Over a nine-month tracking period, it generated a total track of about 12,000 km.

While the Emperor and King Penguin chicks make an abrupt transition to total independence on entering the water, another species breaks the leash to its parents more circumspectly. Gentoo Penguin chicks take their first trip to sea at about 70 days, before finally departing the colony 12 days later. Over the course of this transition period, individual chicks made an average of five trips to sea, returning to the colony to cadge a meal from their parents.[4]

It would be wrong to think that swimming away from the colony precludes a prolonged association between parent and chick. Amongst the smallest auks are the five North Pacific murrelet species in the tongue-twisting genus *Synthliboramphus*, meaning compressed beak. These birds, weighing around 150–200 g, breed in colonies where the eggs - normally two - are laid either in rock crevices at more open colonies, or amongst the tree roots of dank coastal forests, gloomy between the trunks of spruce and hemlock. Parents come and go under cover of darkness, presumably to avoid predators. For the same reason the chicks are raised entirely at sea: they are never fed at the nest. Instead the young, whose legs and feet are almost adult-sized when they hatch from remarkably large eggs*, head to sea a couple of days after hatching.

* The egg mass is about 22 percent of the mother's.

One hundred years ago, George Willett described the scene on a June night at an Ancient Murrelet colony on the Alaskan island of Forrester.[5] "The old bird precedes the young to the water, generally keeping from twenty to one hundred feet ahead of it. A continuous communication is maintained between the two, the frequent cheeps of the young being answered by the parent. . . . The chicks come tumbling down the hill-sides, falling over rocks and logs and, directed by the adult, generally make their way to the bottom of the nearest ravine which they follow to the salt water. Arriving at the water's edge, in response to the anxious calls of the parent who is already some distance out on the water, the chick plunges in and swims boldly out through the surf and joins its parent."

Encouraged and guided by their parents' calls, which they recognize, the chicks swim towards their mother and father who keep moving further out throughout the night.[6] Within 18 hours of departure, the family can be up to 60 km from the nest, as revealed by traditional radio-tracking. Thereafter it appears the family foursome remains together for some six weeks. Unfortunately, miniaturisation has not yet reached the stage where details of that time together, and subsequent family break-up can be followed.

More, information is available for the larger guillemots and Razorbills where the single chick is accompanied by its father for the first two months of sea-going life. Once the fledging chick has survived the potential hazards of gulls, skuas, foxes, and crashes, it makes an offshore rendezvous with the waiting father. Together they swim away, often to nursery areas where the male moults during the period of continuing association.

Thanks to attached geolocators to record position and pressure sensors to record dives,* Chantelle Burke of Newfoundland's Memorial University monitored male and female Common Guillemot activity during

* The devices were retrieved the following spring back at the study colonies, Gull and Funk Islands off Newfoundland. The latter is famous as a nesting site of the lamented extinct Great Auk which may well have had similar breeding arrangements, the father accompanying the chick to sea. Unfortunately there are no contemporary written records to confirm this.

this period.[7] Males made almost twice as many dives as females, on average 104 versus 57 per day, a measure of the extra work involved in finding the extra food needed by the growing chick. In addition the hard-working males tended to dive deeper (49 m) than females (36 m). Then, just over two months after fledging, male and female foraging effort equalised, presumably because the young bird was thereafter able to look after itself.

Why responsibility for the chick devolves solely to the male at this stage remains a mystery. After the female has necessarily done all the egg-laying, it does mean that the total contributions of mother and father to rearing their chick become more equal – but this explanation seems a tad vacuous.

Not all guillemots and Razorbills spend these months of father-chick togetherness in a single area. As the northern winter approaches, there may be advantages in heading south. This is what Brünnich's Guillemots from south-west Greenland do. Rather than make the journey by flying, which is energetically exhausting for the small-winged males and downright impossible for the chicks, they swim south together. Imagine the difficulties for small birds of keeping in contact on such journey when a one metre swell will be towering and a 10 metre swell beyond towering. Again using a combination of geolocators and immersion recorders, Jannie Linnebjerg of Lund University found that male Brünnich's Guillemot parents and their chicks achieved the autumn journey southward of almost 3,000 km entirely by swimming. Further highlighting the energetic advantages of swimming, females and non-breeding males flew only the first 800 km of the journey. The remaining 2,000 km were swum.

Other seabirds might aspire to the aerial route to departure and independence, only to fail in their aspiration. Black-footed Albatrosses leaving their colonies in the north-western Hawaiian Islands are fully feathered and potentially able to fly but, if they falter and land in the turquoise lagoons surrounding the colonies, they risk becoming lunch for Tiger Sharks. Further north in the Pacific, young Short-tailed Albatrosses take time to take wing. After leaving the colony, the fledglings typically make short flights and mostly 'drift' on the sea at walking speed (< 5 km/h) for an average of nine days. Only after this period do they begin sustained flight, indicated by ground speeds over 20 km/h averaged across the 24-hour day.

While these examples suggest acquiring flight skills needs a few days, other species apparently take longer. Leading a Japanese team, Ken Yoda hand-raised a couple of hatchling Brown Boobies which were prosaically named A1 and A2.[8] An acceleration data logger was attached to each bird at fledging to allow precise quantification of the amount of the time spent flapping, gliding and resting. For the post-fledging month of the study, the duration of foraging trips and proportion of time spent gliding during flight increased with the number of days since fledging, whereas the proportion of time spent in flight decreased. It looks as if the Brown Boobies took at least a month to acquire passable flight skills, in particular to shift from energy-demanding flapping to more energy-efficient gliding. This somewhat protracted shift could partly explain why Brown Boobies continue to receive food from their parents for a long time, 17 weeks on average, after fledging.

Without question, successful transition to full independence needs not only competent flying ability but also decent feeding skills. To some extent the two will develop together. A specialised aerial feeding technique is likely to require adept flying. On that basis, we might anticipate that young terns which specialise in snatching fish and other prey from the water surface might metaphorically hang onto the parental (coat)tails for a significant period. Exactly this happens, and it is by no means unusual for juvenile terns to remain with their parents for months. For example, after breeding in Washington state in the northwestern United States, Caspian Terns, a large species with a gigantic orange bill resembling a carrot, head towards the warmth to winter as far south into the tropics as Acapulco, Mexico (17°N). Parent terns carrying solar-powered satellite tags continue to care for the summer's offspring well into the winter.*

Even more prolonged parental feeding of fledglings is the habit of frigatebirds. Henri Weimerskirch, an urbane French research scientist,

* Considering all the tern species, this is probably one of the longest periods of association between parents and their flying offspring. Another species with a notably long period of association is the Black-naped Tern, where parents and young can remain together for up to 180 days.

has unquestionably been at the forefront of modern pioneering seabird research. As the years have passed and the sinews stiffened, Weimerskirch has shrewdly shifted his own research from the petrels of the Southern Ocean towards the frigatebirds of the tropics. Accordingly his clothing has shifted from parka to floral shirt. At the 2015 2nd World Seabird Conference in Cape Town, he described how Great Frigatebirds undertake a protracted transition to independence, competent flight, and specialised feeding which may involve catching flying fish at the sea's surface or harrying other seabirds into disgorging their food in flight, and then catching the vomited meal even before it has fallen to earth or water.

The juvenile Great Frigatebirds hailed from Europa, a low atoll between Mozambique and Madagascar. For the first six months of flying life, the juveniles go to sea by day but return to land by night to be fed, normally by their mothers. Then the satellite-tracked young birds move north up the Somali coast, perhaps looking down on the region's contemporary human pirates. The journeys are relaxed, around 450 km/day. The birds alternate periods of soaring in circles, attaining heights up to 3,000 m, and periods of slow descent. While soaring, the frigatebirds do not flap their wings but rely on differences in air speed between different blocks of air to gain height, so-called dynamic soaring. Using this tactic, the young frigatebird remains in flight for up to two months at a stretch,* sometimes even passing close to but not making landfall on the scattered islands of the Indian Ocean. Only occasionally is the pattern broken with land-based rests of a day or so on isolated islets of the Seychelles or Chagos archipelagoes.[9] As the young birds pursue repeated clockwise circuits of the Doldrums of the central Indian Ocean for a year or more, it is an immensely leisurely entrée to independent life.

Some young seabirds evidently take weeks or months to break the link with their parents. This may be due in part to the time taken to hone feeding skills, a process potentially taking months or years, as will be explored further in the next chapter. But the acquisition of flying skills is likewise not trivial. On the other hand, the very rapid, very long-distance journeys to be described later in this chapter suggest, for some species, those flying skills are adequate or better from the get-go.

* For eight juvenile birds, the average maximum time spent aloft was 41 days.

Shortly after becoming independent of its mother,
a juvenile Great Frigatebird may cruise the Indian Ocean,
remaining continuously in flight for up to two months.

Once a seabird has the flying powers to quit the immediate nesting area, it is faced with a choice. One possibility is to explore the local region, loosely interpreted. This tactic potentially tells the explorer about the whereabouts of the best feeding areas. Presumably remaining with parents facilitates this process. However, even in the absence of parents, following members of one's own species could be a useful pointer to feeding zones and migratory routes.

Fifty years ago, Mike Harris, working on the Welsh island of Skokholm, swapped eggs between the nests of Herring Gulls which routinely remain close to the colony through the winter, and the nests of Lesser Black-blacked Gulls which, at least in the 1960s, then spent the winter in Iberia.[10] The young Lesser Black-backs were barely affected by the treatment; they still went to Iberia. But the young Herring Gulls, raised by migratory Lesser Black-backed foster parents, moved further south through France than 'normal' young Herring Gulls. Carried out before the advent of modern electronics, and indeed around the time man first stepped on the moon, these results could indicate that young birds follow the species they perceive as their own as a means of finding their way in the world.

There is no reason why exploration of new areas should cease after the bird's first few months. Børge Moe, from the Norwegian Institute of Nature Research, told the 2nd World Seabird Conference how young Black-legged Kittiwakes, tracked with geolocators from their home colony in Kongsfjord, Svalbard, wandered far and wide in the years between fledging and recruitment to the colony 2–4 years later. For example, in their first summer, around one-third of these small gulls with a wafting flight ventured into northern Baffin Bay, surrounded by Ellesmere Island and north-east Greenland. After a winter passed in the grumpy lumpy seas east of Labrador and south of Greenland, where the adults also winter, the Kittiwakes spent the next summer back closer to home, at Svalbard. In contrast other immature Kittiwakes used their first summer to explore east from Svalbard towards the Russian islands of Novaya Zemlya, a site of much Soviet nuclear testing. Whether these explorations ever led to the birds settling away from their colony of birth is not known – but it seems plausible.

If the seasons are changing fast – perhaps equinoctial gales threaten – an alternative tactic is for the fledgling to depart smartly for the non-breeding area, possibly thousands of kilometres away. Traditional observation has left no doubt that these massive journeys, sometimes between hemispheres, can be impressively rapid. Famously, a young Manx Shearwater, ringed at a Welsh colony, was found on a Brazilian shore 16 days after ringing. Since the finder thought it had been dead at least three days, the 9,600 km journey had been completed at 740 km/day, perhaps faster. No flying lessons required there! Lockley[11] was correct to wonder whether the Manx Shearwater "might be a great annihilator of distances". And the very direct journey, in the absence of parents, suggests route choice was genetically determined.

Or take the case of the Short-tailed Shearwater, an extremely abundant breeder and the harvested muttonbird of islands around south-east Australia, including Tasmania. Immatures leave the breeding grounds in March, breeders follow in mid-April and fledglings make up the rearguard in late April–early May. The trans-equatorial movement north of perhaps 30 million of these shearwaters is surely one of the world's greatest bird migrations, a fluttering avalanche of 20,000 tonnes of sentient flesh, roughly half the weight of the Titanic. This torrent's track is initially southward and eastward, thus avoiding the coast of Australia. Then the birds turn north-west, passing both east and west of Fiji, and probably making fast progress across the tropics. If travelling at 50 km/h for 20 hours a day, progress could be around 1,000 km/day. Birds reach the Sea of Okhotsk and the west end of the Aleutian chain in late April. Huge numbers spend the northern summer in the Bering Sea.*

The picture just painted covers the mass of migrating shearwaters. It does not specifically focus on the fledglings, and it remains troublesome to gather data on the long-distance movements of smaller species, like shearwaters, on their first migration. Because, even if they survive, the birds will not return to the colony for several years, so the chance of capturing them to retrieve a geolocator or GPS device is small. Satellite tags are a better option, but are heavier. Nonetheless a satellite study of Scopoli's Shearwaters was able to assess the speed of fledglings starting independent life and heading from their breeding colony off Marseille towards the Strait of Gibraltar, and then south to the wintering area off west Africa. Within the Mediterranean the youngsters and older birds, both immatures and adults, travelled at about the same speed of 90 km/day. On reaching the Atlantic the pace picked up and the fledglings (240 km/day) travelled faster than the older birds (177 km/day). However this shearwater study by Clara Péron and David Grémillet remains an exception.[12] The best information about those first long airborne journeys comes from larger species, especially albatrosses.

Working on the sub-Antarctic Iles Crozet, Henri Weimerskirch and his team tracked 13 juvenile Wandering Albatrosses by satellite telemetry during their first year at sea.[13] Leaving the island, the birds first spent about five days drifting on the water. As soon as a southerly or south-westerly wind started to blow, they took to the air and headed between north and north-east. So consistent was this route choice that it was likely genetically determined. Then, fairly quickly, the juveniles upped their daily travel distance to some 600 km. This is the same distance adults routinely cover in a day and was very probably linked to the attainment

* The relationship between this curving trans-Equatorial track and the prevailing winds is treated in Chapter 6.

of comparable flight efficiency. Six months after quitting Crozet, the birds were concentrated in subtropical waters just south of Australia and eastward into the Tasman Sea, with the young males possibly going further east than the females.[14] The subtropical waters used at this phase of life were distinctly north of the zones used by older Wanderers. Remarkably, the estimated average distance covered during the first year at sea was 184,000 km, equivalent to four times round the world.

Wandering Albatrosses are not the only species where the young birds have different target destinations to their elders. A similar pattern has been discovered in Black-footed Albatrosses tracked from Midway Atoll.[15] Situated at the north-western end of the Hawaiian chain and, appropriately, about half way across the Pacific, the island saw a key battle between Japanese and US forces in 1942. Today around one-third of the world population of this albatross species breeds on the atoll. After a few days bobbing on calm waters within the atoll, and risking Tiger Shark attack, the fledglings head northwards for some 800 km. At this point the birds are nearing the more productive waters associated with the so-called North Pacific Transition Zone some 10 degrees north of Midway Atoll. But, perhaps unexpectedly, the young actually linger south of this zone, remaining within the gyre circulating clockwise around the middle latitudes of the North Pacific. Occupying the centre of this clockwise gyre lies the so-called Great Pacific Garbage Patch. At least the size of Texas, the vortex traps plastic and all manner of garbage. But why should the young albatrosses elect to remain here rather than join the adults further north in the productive waters off northern Japan and off the Aleutians? Perhaps it is the very presence of those more experienced and competitive adults that is the problem. They might be simply better at catching food than the juveniles, and oust the young from the best patches. Perhaps the fact that, unlike the adults which moult after breeding, the young do not have to face the extra energetic costs of moulting, means they can balance their daily energy budgets in less productive waters. The situation mirrors that of the Crozet wanderers where, in the months after breeding, the young are spatially separated from the adults.

The north/south split seen in the wanderers, young to the north, adults further south, has also been discovered in another species of the

Southern Ocean, the Southern Giant Petrel. Giant petrels have a fierce hooked bill that can tear open the belly of a seal or penguin. Pair that bill with a blood-smeared head from which a pale maggot-coloured eye peers, and it is a rare sentence that contains both 'giant petrel' and 'cute'. Adult Southern Giant Petrels from the Crozets tend to remain in the southern Indian Ocean outside the breeding season, while juveniles are not only consistently further north but disperse all around the globe in those southern latitudes.[16] Indeed, so extreme is the nomadism of the juvenile Southern Giant Petrels, that all those studied circumnavigated Antarctica within three months of leaving their colony.[17] But, importantly, their more northern distribution puts the juveniles at far greater risk of encountering a hook streaming behind a tuna long-lining vessel, of which more anon.

An equally extreme separation is achieved, albeit rather differently, by the Southern Giant Petrels from a colony in Patagonian Argentina. Adults persistently use the Argentinian shelf while, within a month of fledging, young birds have entirely quit the zone of adult activity by moving northward as far as waters off southern Brazil.[18]

While the juvenile Wanderers head east from the French sub-Antarctic islands, juveniles of two other species breeding on the Crozets, White-chinned Petrels and Sooty Albatrosses, head north and west to the subtropical waters of the Benguela and Agulhas Currents, respectively west and east of South Africa. It is very evident that, even at this early stage in their lives, the young of different species have different target destinations.

Indeed, the young of the same species from separate colonies can head to different destinations, even when those colonies are separated by barely half a day's flying, at albatross speed. All the world's Shy Albatrosses breed at just three islands off Tasmania, namely Albatross Island, Pedra Branca and the Mewstone. When monitored by Rachael Alderman's team,[19] the tracked fledglings from the first two sites remained in the Great Australian Bight and barely glided west of the Western/South Australia border in their first three months of independence. In contrast, Mewstone birds had a hotspot of activity off Albany at the extreme south-west of Australia.

When young seabirds first dip a toe in salty water, it is sometimes a rather tentative dip. Think of the Gentoo Penguins returning repeatedly to the colony to be fed, or a young gull, able to fly, but still mewing plaintively and complaining at its parents' lack of due feeding diligence. Sometimes the transition to independence can be prolonged because parents and chicks remain together whilst at sea, for weeks or even months. But sometimes the youngster quits the colony by itself. With abrupt certainty it must adjust from life on land, assured of regular feeding visits from its parents, to independent life at sea. Following that adjustment, the young bird (depending on species) may not set foot on land for several years. It embarks on a period of footloose immaturity before it joins a breeding colony.

CHAPTER 3

# The Meandering Years of Immaturity

Among humans, the teenage years are a time of trial and, all too often, error. This is a period when young people take risks, ignore warnings and, perhaps, undergo a period of accelerated brain development. In a nutshell, it is a period of exploration. Whilst the analogy would soon fracture if extended too literally to seabirds, a seabird's immature years are also a period of learning and exploration.

Having fledged, a young seabird embarks on a period without breeding responsibilities. No species breeds when less than two years old, and some albatrosses do not breed until ten or older. Although certain species, such as gulls and cormorants, may routinely visit land during this period, others, for example petrels and penguins, can spend several years without ever putting webbed foot on *terra firma*. Then, as the years pass and the bird matures, colonies are visited, and one colony is eventually chosen as the breeding place. A mate is found and breeding commences. This extended pre-breeding period has long puzzled biologists, especially as some species seem physiologically able to breed years before they do so. For example, a team from the British Antarctic Survey peered laproscopically at the testes of five-year-old Wandering Albatrosses on Bird Island, South Georgia. Despite this species rarely breeding before the age of eight, the testes of these non-breeding males aged five or older were as large as those of many breeding males, and levels of testosterone in blood samples were in line with the range of breeding males.[1] Why do they delay so long?

The answer could be linked to the risks and benefits of breeding. There is likely to be a risk of predation when visiting the colony, there could be risks associated with flying substantial distances between the colony and the prime feeding regions, and parenthood itself may be physiologically exhausting. On the other hand the benefits, measured in terms of fledglings hatched and despatched, hinge *inter alia* on the bird knowing where to find food and catch it efficiently. Possibly several years are needed before this know-how becomes sufficiently refined for the potential benefits of breeding to outweigh the likely risks.

This life-history pattern means there is a time gap between a bird's first flight and its return to the colony months or, more likely, years later. Even with the availability of modern devices, this poses problems for the researcher seeking to track a bird's location and activity. The batteries of devices deployed at fledging may expire. Gadgets attached to the plumage will fall off when the bird moults. And, of course, the bird has to survive those immature years if its device is ever to be retrieved. No wonder scientists refer slightly forlornly to this period as the 'lost years'. Yet it is also the period when the seabird is likely to explore the oceans, discover the best feeding areas, and perhaps develop the individually-consistent habits that are a feature of so many adult seabirds, a fascinating topic to which I shall return in Chapter 7. Without better understanding of the pressures on seabirds during those missing years, our ability to improve their lot is severely compromised. The urgency is only compounded by the observation that, in species with a substantially delayed start to breeding, over half of all individuals alive may be in their pre-breeding years.

In the previous chapter, I recounted how Black-legged Kittiwakes fledging from Svalbard variously went, in their early years, to the northern regions of Baffin Bay to the west and to Novaya Zemlya to the east. They may even have visited other colonies during those early wanderings. Børge Moe's study hinged on the recovery of geolocators up to four years after deployment. Such long-term deployment and eventual retrieval is currently exceptional. Probably this will change as battery lives are extended.

Instead of following birds from fledging, an alternative means of checking behaviour during the years of immaturity is to attach devices to young birds as soon as they (dare) set foot in a colony. This may be less than ideal; the birds have been at sea, their whereabouts unknown, for several gap years. Nonetheless, it potentially allows the checking of a period in their lives when, following the arguments sketched above, they may still be behaving like teenagers, exploring, perhaps visiting other colonies, and improving their feeding skills.

There are three principal means of identifying these young birds at a colony. The most conclusive is the presence of a metal ring applied to the leg when they fledged. This has the great advantage that it means the age of the bird is precisely known. Another possibility, available only for a minority of species whose feather patterns change with age, is to use the plumage pattern as a tolerably accurate signal of age. For example, this works for Northern Gannets and Wandering Albatrosses and other great albatrosses in the genus *Diomedea*. Finally the behaviour of immatures is different from those birds actively breeding. The former may spend time in groups, so-called 'clubs', while the latter scurry like responsible employees between home and hearth and their workplace, the sea. The risk here is that birds of breeding age, taking a year off for whatever reason, may mistakenly be included in the sample of putative immatures.

While most of the information I am about to describe comes from electronic devices, there is another important, slightly indirect, route to knowledge. Stable isotopes [see Box], extracted from minute samples of a bird's blood, muscles, bones or feathers, can provide supplementary information on its diet and travels.

## STABLE ISOTOPES

Many chemical elements naturally occur in more than one form. All forms (both stable and the less stable, radioactive isotopes) have the same number of protons but differ in the number of neutrons in the nucleus. For example, both Carbon 12 ($^{12}C$) and Carbon 13 ($^{13}C$) have six protons per nucleus but differ in the number of neutrons, six and seven respectively. Carbon 12 is around 100 times more common than Carbon 13 in nature. A 50 kg human contains about 11.4 kg of Carbon 12 and 140 g of Carbon 13.

Various stable isotopes have proven useful in palaeontology, archaeology and ecology.[2] One reason for their utility is the fact that the different isotopes, with their different atomic masses, behave differently as they pass through natural systems, and become incorporated in animal tissues. For example, the heavier isotope of nitrogen, Nitrogen 15, tends to be retained by organisms while the lighter 'regular' isotope, Nitrogen 14, is excreted. Thus, when one organism eats another, Nitrogen 15 becomes more concentrated in the animal doing the eating. As this isotope accumulates further up the food chain, so an animal's nitrogen isotopic signature, ascertained via mass spectrometry, is an indicator of its diet, be it a top predator, an animal that eats grazers, or a grazer itself.

Not only do isotopes differ in how they are processed by animals and plants, they can also show broad geographic patterns. For example there are changes with latitude in the sea in both the $^{13}C/^{12}C$ and the $^{15}N/^{14}N$ ratios. On land, the ratio of the hydrogen isotopes $^{2}H/^{1}H$ tends to decrease at higher latitudes and altitudes ($^{2}H$ is deuterium), while there is a complex interaction between latitude, altitude and rainfall on the ratio of the oxygen isotopes $^{18}O/^{16}O$. Both $^{15}N/^{14}N$ and $^{34}S/^{32}S$ (sulphur isotopes) ratios are higher in the sea than on land. Potentially therefore, the stable isotopes prevailing in the environment where a bird turns food into body tissues, be it new flesh or new feathers, can leave an informative signature. This signature enables researchers to understand where in the world the new tissues were manufactured.

In the case of seabird studies, stable isotopes have been used to shed light on diet and roughly where a bird was when new tissues were grown. The accuracy of that geographical information may be no better than a few thousand kilometres, but it is better than no information at all, especially when it can be gleaned from a tiny blood or feather sample, all that is required for successful mass spectrometry. Therefore the technique has the ethical and scientific merits that it is harmless and can potentially be repeated several times during a bird's life.

First catch your immature seabird. I myself have caught many thousands of seabirds by metal hook or nylon noose around the leg, by fleyg (effectively a giant lacrosse net used especially by Faroese and Icelanders to catch Atlantic Puffins for the pot), by mist net and, of course, by a straightforward grab. Steve Votier, an ornithologist at the University of Exeter, has a personal website that shows one of the other tricks of the seabird trade, catching gannets by using a long pole to noose them around the neck.[3] Using this technique, his team noosed gannets at the world's third largest colony on Grassholm, off south-west Wales. Each of the captured birds, five immatures and 25 adults with chicks, was released after the brief indignity, with a satellite tag, programmed to obtain an hourly-GPS fix. The contrast in the tracks of the two groups was striking. The average out-and-back journey of an adult was 370 km. None went north into the Irish Sea and the furthest west reached by any bird was south-west Ireland. The immatures travelled three times the distance on their excursions from Grassholm, and showed considerable variety in their tracks. Two rounded Brittany to enter the Bay of Biscay while another once did a circuit of Ireland and once ventured among the Scottish Hebrides.[4] (See Map 1.)

Not only did the tracks of the immatures tell of exploratory behaviour, so did the fact that two of the five visited other gannet colonies on their excursions. One bird called at no less than three Irish colonies while another made a prospecting visit to the Rouzic colony off the northern coast of Brittany. It would be fascinating to know more about whether and in what circumstances these pre-breeding visits led to eventual settlement as a breeder in later years.

Although the picture was slightly clouded by technical issues, Votier's analysis of the stable isotopes in the gannets' blood suggested that, compared to adults, the young birds were either eating larger fish or fishery discards, the fish caught but not brought ashore by fishers. It is at least a possibility that they were feeding in different areas. Not only is that compatible with the different tracks, but it would also enable the younger gannets to avoid feeding alongside the potentially more skilled adults.

Another study, this time in the turquoise Mediterranean rather than the drabber Irish Sea, matched Votier's work in showing that immature seabirds may visit several colonies. In the latter part of the breeding season, namely late August and early September, Clara Péron and David Grémillet attached satellite transmitters to both immature (aged 4–6) and breeding adult Scopoli's Shearwaters at a colony at the southern end of Corsica.[5] In the next month adults concentrated their foraging in waters just west of Corsica and Sardinia, while immatures often headed east towards mainland Italy and there passed within 5 km of other shearwater colonies. Such 'prospecting' visits mostly happened by night, presumably when this nocturnal bird would gain the most useful information about the merits of the colony as a possible future home.

Let us ruminate for a moment, and set aside the possible advantages to the young bird of scouting for other colonies at which it might eventually settle. If younger and older birds are feeding in different areas and possibly catching different prey, that could be for one or both of two reasons. Here is the first. Given that the adult birds are likely to need to visit the colony more often, perhaps to feed the chick, then the advantage of feeding close to the colony may be greater for them than for the younger birds. The latter then benefit from feeding elsewhere in a less crowded sea area. That would be analogous to those occasions when I have abandoned a crowded payment till close to the entrance of a department store, and sought a less crowded till on another floor. The walk to the other floor was a trivial price to pay for quicker service.

This rationale for separation between younger and older birds is conceptually distinct from the second. If younger birds are less efficient at feeding, then they may benefit from keeping their distance from the older birds, regardless of where the two groups feed. And we have already posited that the delayed start to breeding of seabirds may be associated with a prolonged improvement in feedings skills.

A nice example of birds of different ages feeding in different areas, coupled with evidence that the young are less efficient at foraging, also comes from the Irish Sea. Annette Fayet and a team from the OxNav group – the notably straightforward name for scientists at Oxford University studying animal navigation – focussed on Manx Shearwaters. They breed on Skomer Island, within sight of the Grassholm gannets.[6] No nooses were needed to catch these birds since immatures can be plucked from the ground at night, when they visit the colony. Adults too are simply captured, by withdrawing them from their nesting burrow. At the season of this project, incubation was nearing its end and small chicks were hatching. Having attached GPS devices to 50 immatures and 27 adults, Fayet then faced the need to recapture the birds to retrieve the devices. That was straightforward for the adults; the birds returned to their marked burrow. For the immatures, it was anything but straightforward. Imagine a pitch dark June night, with tens of thousands of birds flying, calling, and landing with a thump. Which is the one with the GPS device? To increase her chance of finding it, Fayet stuck reflective tape on the GPS loggers, and also configured the loggers to emit either a blue flashing light or a radio signal which alerted two base stations, and hence the researchers, that a target bird was in the vicinity that night.

Fayet summarizes the frustrations of her fieldwork, "We hid on the colony dressed in dark clothing, armed with night-vision scopes and walkie-talkies, slowly crawling towards birds to catch them by surprise when they eventually returned to the colony. By the end I had spent so much time desperately looking for blue flashes in the dark that I ended up seeing them everywhere!"

Perseverance was rewarded by useful data from 20 immatures and 19 adults - and the results were striking.

The immatures made shorter trips than the adults, both in terms of duration and maximum distance from the colony (over 200 km for adults, an average of 135 km for immatures), and those trips took the birds in markedly different directions. While the immatures concentrated south of Skomer, adults either headed west to waters off southwest Ireland or north to the northern Irish Sea. (See Map 2.) As with gannets, the Manx Shearwaters of different ages were spatially segregated. But the icing on the study's cake came when Annette Fayet used the pattern of GPS fixes to determine for how long on each trip a bird was feeding. Roughly speaking, the bird was reckoned to be feeding when it was moving relatively slowly over the sea, and turning often. This enabled her to calculate how much mass was gained per unit time spent feeding. It was significantly less for the immature birds. They appeared to be less successful foragers, even when feeding away from any possible interference from adults. What remained unresolved was the

It is pitch-black on Skomer Island off the coast of Wales. Clamorous Manx Shearwaters are present in their thousands. The challenge for researchers: find the handful of birds carrying GPS devices.

extent to which this inferior performance of the young birds was due to lower efficiency, and to what extent it was because their southward track from Skomer took them to waters less productive than those used by the adults.

A picture is emerging of immature and breeding seabirds using different sea areas, possibly in part because the young birds are less efficient foragers. If correct, we might expect that, through the several years of immaturity, the young bird's behaviour at sea would progressively converge on that of the adults. An example where this might be so is provided by the Wandering Albatrosses tracked from the colonies on Iles Kerguelen and the Crozets in the Southern Ocean.[7] In their first five months after fledging, young Wanderers are as well able to take advantage of tail and side winds as adults. This aspect of flight apparently needs little learning. However they spend progressively less of their time on the water and more on the wing, covering greater daily distances. This is almost certainly associated with developing the flying skills described in Chapter 2. Then, following that tiresome gap when the birds are beyond reach, immatures between four and eight years of age reverse one trend but continue another. They spend a little more time on the water than five-month juveniles but travel further. Have they now mastered flight? Finally, adults spend more time on the water and cover fewer kilometres per day than the immatures. Perhaps the adults refined their fish- and squid-snatching skills to such an extent that they can fulfil their daily energy needs with less effort.

Completely compatible with the argument that seabirds begin breeding when the benefits exceed the costs is the possibility that feeding skills may continue to improve, even after first breeding. Such has been found in Cory's Shearwaters, a species named after a wealthy American, Charles Barney Cory, who was once the golfing champion of Florida. The shearwaters breed on islands of the north-east Atlantic, including the small Berlengas archipelago off Portugal. Here, almost 40 male shearwaters were tracked with GPS loggers in the period immediately before the males' partners laid. The males were split into two groups, the inexperienced with a record of two or fewer breeding successes and the successful with more than two successes. The experienced males consistently concentrated in productive waters close to the Portuguese coast while the inexperienced roamed more widely into less productive seas further offshore, especially in the years when overall conditions were probably poorer.[8] Perhaps these positional differences, supported by stable isotope differences between the two groups, arise because the younger breeders are still honing their feeding skills and/or learning the likely whereabouts of a shearwater's supper.

The age-related changes in foraging areas used do not stop after the early breeding years. I have already mentioned how juvenile Wandering Albatrosses operate north of older birds. Remarkably, there are continuing shifts among breeding birds, principally among the males. In a paper aptly titled 'Lifetime foraging patterns of the Wandering Albatross: Life on the move!', Henri Weimerskirch[9] details how the latitudes used by female albatrosses, when on feeding trips during incubation, shift little with age. They remain around 40°S. Males visit similar latitudes to females in their early breeding years, say at 10 years old. Thereafter they steadily shift southward, in marked contrast to females. By the time he is 40, becoming ever whiter of plumage, and showing signs of senility (and therefore lower annual survival), a male's trip is likely to take him south to the icebergs, around 60°S. A direct consequence of this age- and gender-related effect is that the stable isotopes measured in the two sexes increasingly diverge as they age, from no differences in immature birds to major differences in old (>30) birds, reflecting the males' use of colder seas. Exactly why the males move southward with age remains a puzzle.[10] A dismal explanation is that they are forced south by competition with more vigorous males still in their twenties. A more positive explanation is that they benefit from the stronger winds, reducing travel costs in those higher latitudes.

Many of the details of birds' activities in those early years before they re-appear at a breeding colony remain sketchy. The omissions will doubtless be somewhat rectified as batteries for geolocators become longer lasting, and the use of long-lasting solar-powered PTTs becomes more widespread. But it will certainly remain challenging to document the daily routine of an animal that has no reason whatsoever to come within 100 kilometres of an eager researcher. Only when the bird needs to make landfall to breed does the advantage shift towards the researchers. They will be waiting - and the findings already secured are astounding.

CHAPTER 4

# Adult Migrations

20,000 Leagues over the Sea

Sometimes our knowledge of the natural world relies on deduction rather than direct observation. When a Sperm Whale dies and is autopsied, its stomach is full of the hard indigestible beaks of squid. Now imagine watching a Sperm Whale off the Kaikoura coast of New Zealand. Its flanks are pocked by circular marks, presumably the battle scars left by past encounters with the suckers on the tentacles of Giant Squid. So when the whale flicks its tail skywards, before sliding vertically into the abyss, the 3 km depths of the Kaikoura Canyon, it is impossible to resist the urge to wish it well ahead of the imminent life-and-death combat in the dark under immense pressures that would crush the housing of an everyday underwater camera. Even without photographic or observational back-up, we are ready to believe such battles take place because how else would the whale's stomach come to be filled with squid bits.

Similar deduction has improved knowledge of the extremes of bird migration. As the autumn equinox approaches western Alaska, Bar-tailed Godwits, wading birds standing 30 cm tall on spindly legs, are becoming restless. One night, often when a tail wind from the north is blowing, they leave Alaska for the south. Across the vast swathe of the Pacific Ocean south from Alaska, islands are desperately far apart. Moreover, few Godwits are ever seen on the scattered atolls at this season. Nor are the Alaskan birds, some of which carry leg tags, seen on the continental rim of the Pacific. Yet, late September and early October witness the arrival of the Godwits at New Zealand's estuaries. Is it possible that the birds fly from Alaska to New Zealand in a single 10,000 km flight, losing at least half their take-off weight* en route? Such seemed very likely but it was only when satellite transmitters were attached to the birds in 2006 and 2007 that the non-stop week-long journey was confirmed.[1]

Deduction has also pointed to the likelihood of other remarkable migratory feats. Blackpoll Warblers winter in northern South America and breed across North America from Newfoundland to Alaska. Check an atlas and the obvious southbound route from New England to Colombia or Venezuela is a direct flight across the Gulf of Mexico. Such a feat had long been suspected of these 12 g birds because of radar evidence (echoes from small birds heading seawards at the appropriate season) and because of the scarcity of these warblers along the United States' eastern seaboard anywhere south-west of North Carolina in autumn. Only the fitting of 0.45 g geolocators to 37 Blackpoll Warblers in Vermont and Nova Scotia in the summer of 2013 clinched the issue. Five Warblers were recaptured the following spring and the downloaded

* The birds set forth from Alaska at a mass of 325–400g. About 55 percent of that mass is fat which is probably almost entirely used by landfall.

data showed all had headed south over the ocean on their autumn migration. All had stopped on a Caribbean island (Hispaniola or Puerto Rico or the Turks & Caicos) before completing the journey to continental South America. These tiny birds had completed a trans-oceanic 2,500 km journey lasting up to three days.[2]

Modern devices sometimes confirm what is already suspected of seabird migrations. And sometimes the devices have revealed journeys barely suspected, where the reaction in public might be along the lines 'Well, I wouldn't have guessed that!' and, in private, a more colourful 'Holy S$%t!' Who would have anticipated that the male and female of a single pair of Sabine's Gulls, nesting together in the Canadian Arctic, would elect to spend the northern winter far apart, she over the Humboldt Current off Peru and he over the Benguela Current off South Africa, only to re-unite the following spring back in the north?

In this chapter I will sketch a handful of the more remarkable migratory journeys that have been documented as information has flooded in from modern devices, very often attached to adult seabirds one breeding season and retrieved the following season. Thus the birds' fidelity to a colony renders them far more amenable to research than during those 'lost' years of immaturity. From this research findings have emerged that sometimes startle, sometimes amaze, and sometimes merely plug knowledge gaps. The results have led to researchers then asking, and partly answering, more subtle questions. What are the routes taken between breeding and non-breeding areas? Are those routes traversed at a more or less steady speed, or is the journey completed by a mix of long distance flying days interspersed with feeding 'rests'? Do all individuals of a population use the same non-breeding area? If the answer is no, are individual birds faithful to just one area? Are there any differences between the non-breeding areas or migratory routes used by males and females? Perhaps most intriguing of all, is there any sign that the two pair members stick together outside the breeding season during the rough times and calmer periods at sea?

Anticipated but still breathtaking has been the confirmation of the incessant flying feats of Arctic Terns, each weighing 100–125 g, similar to a small and barely acceptable steak. More poetically these sea swallows shuttle back and forth between far north and far south on a wing and a prayer:

The Arctic Tern's Prayer

Tell the air to hold me in the rushing heart of it
And keep its paths straight
Away from home let there be a land that
Flows with fish and flies
And let it taste like it tasted at home

– *Mary Anne Clark*

With the aid of geolocators, a multinational team secured data on the tracks of 10 birds from Greenland and one from Iceland.[3] Southbound, the terns travelled southwest to a stopover region of deep water in the eastern portion of the Newfoundland Basin and the western slope of the mid-North Atlantic Ridge between 41–53°N and 27–41°W. There they lingered for 3–4 weeks. Continuing southward, all 11 birds headed southeast toward the West African coast. South of the Cape Verde Islands (~10°N), however, migration routes diverged: seven birds continued to fly south parallel to the African coast, whereas four others crossed the Atlantic to follow the east coast of Brazil. (See Map 3.)

Once south of the Cape of Good Hope, there were more east-west movements but still the terns generally continued to push southward and spent the austral summer roughly south of 58°S, at the mouth of the Weddell Sea, where oceanic productivity is high. The S-shaped return in spring followed a different route to that of autumn. Initially the terns flew north, until about 30°S. Then they swung left and followed a north-westerly route across the Equator until reaching tropical waters east of the Caribbean. Then a right turn took them north, or a little east of north, back to Greenland and Iceland. That final leg saw the birds pass back through the area used during the southbound stopover. However, northbound that area apparently held little interest for the migrating terns.

The numbers generated from these travels are exhausting. The journey totalled at least 72,000 km, comprising 35,000 km southbound at about 330 km/day; 11,000 km while (relatively!) dawdling in Antarctic waters, and finally 26,000 km northbound at 520 km/day.

Twitchers, who generate long lists of bird species seen, are notoriously competitive. Scientific ornithologists are less so – or is that a false impression? No sooner had the ink dried on the details of the migrations of the Greenland Arctic Terns than a Dutch group published a paper entitled 'Arctic Terns from The Netherlands migrate record distances across three oceans to Wilkes Land, East Antarctica'.[4] Surely that title carries more than a hint of one-up-man-ship. While the terns' routes through the Atlantic were not dissimilar to those of the Greenland and Iceland birds, the Dutch birds accumulated a total journey distance 20,000 km greater, about 90,000 km, because they wintered well to the east of the Weddell Sea off Wilkes Land, which is approximately south of India. It seems only a matter of time before a study carried out on Arctic Terns nesting in north-west Russia, and having to round northern Scandinavia to reach the Atlantic, provides the world with the first 100,000 km tern journey.

As with the terns where traditional study techniques had given strong indications of the likely travels, so too with Grey-headed Albatrosses. With a head of soothing pearl grey, the species nests at a handful of localities deep into the Southern Ocean, mostly between 45°S and 55°S. This is a bird that lives life in the relatively slow lane. If successful in rearing a chick, it takes a sabbatical year away from parental duties and is absent from the colony for around 18 months before starting again to breed. Since the species is seen all around Antarctica, there was every possibility that birds could complete one or more circuits of the continent in their sabbatical year. It fell to workers from the British Antarctic Survey, pioneers in the development of geolocators, to test the possibility, and it was Grey-headed Albatrosses from the colony on Bird Island, South Georgia, that were the subjects. The albatrosses partly confirmed and partly confounded expectations.[5]

Geolocators retrieved from 22 birds revealed that the peregrinations in the 18-month sabbatical period fell into one of three groups. Some albatrosses remained in the south-west Atlantic, using the same marine areas as South Georgia breeding birds. Some headed east to spend the southern winter in the south-west Indian Ocean, possibly returning to the South Georgia region in the summer. And, thirdly, 12 birds completed at least one eastward circumnavigation, while three of them undertook a second circumpolar trip during their absence from Bird Island. All the circumnavigators went eastward, propelled by the westerlies of these latitudes. The poet might wonder about the joy of opening one's wings and sliding downwind for a full circuit of the world we inhabit. The scientist might wonder about the relative merits of these distinct sabbatical strategies. Both could agree that the fastest circumnavigation of 46 days, including spells averaging 950 km/day was stupendous. Even modern yachts, able to surf the Southern Ocean's swells at 35 terrifying knots, would be hard pressed to keep pace with a Grey-headed Albatross.

While the tern and albatross studies broadly confirmed pre-existing knowledge, other studies have eliminated a total absence of knowledge. The Atlantic Puffin with its colourful bill and comic mien breeds abundantly on Britain's offshore islands. It is a favourite among day-trippers to accessible islands such as Skomer and the Farnes. Yet Puffins are ashore only between April and August. They all but disappear in the intervening months. The OxNav group, led by Tim Guilford, has shed light on the winter travels of the Skomer Puffins.[6] In August, most birds migrate away from the colony, most in a westerly or north-westerly direction, some as far as Greenland, some more locally, whilst some move southwards towards France and Biscay. In autumn, mostly October, all move northwards or north-eastwards into the North Atlantic. Then, later in the winter, they travel southwards, some as far as the Mediterranean, before returning (from a variety of directions) to the colony in spring.

Given these ramblings, it is not surprising that the Puffins' winter whereabouts were unknown. However they are not random ramblings; individual birds take remarkably similar routes in successive years (see page 136). Moreover, a similar study of the Puffins sharing Skellig Michael off west Ireland* with the island's extraordinary monastic ruins yielded comparable findings.[7] After breeding the Puffins headed west. By September, some birds were off the coasts of Newfoundland and

* Skellig Michael is the location of the final scene in *Star Wars: The Force Awakens*.

Labrador, possibly enjoying the same food as their fellow Puffins breeding rather later in North America. Remarkably, having made the month-long westward journey across the Atlantic, the Irish Puffins lingered only 2–6 weeks before commencing the leisurely return eastwards which, as with the Skomer birds, often took them initially northwards into the Denmark Strait and Icelandic waters. The reason why this pattern should be shared by the Puffins of Wales and Ireland remains unknown.

Another auk has been discovered to undertake a trans-oceanic east-west migration after breeding – but this time across the Pacific. Ancient Murrelets breeding at the major colonies of Haida Gwaii (formerly known as the Queen Charlotte Islands and the Charlottes) were given geolocators by a team led by Tony Gaston. Instead of heading south after breeding from British Columbia to coasts off the western United States, the Murrelets went west, so far west that, by January, they were in seas between Korea and Japan. The round trip, 8,000 km in each direction, represents the longest migration of any auk. No other North Pacific species undertakes a similar journey, nor, Gaston writes, "is there any evidence that the wintering area presents unusually rich feeding opportunities". That, of course, is the scientific way of saying he has no idea why the birds undertake this epic migration.

Like the Puffin and the Ancient Murrelet, the Common Diving Petrel is not an obvious long-distance traveller. With stubby wings for both flight and underwater propulsion and a chunky body, its shape when airborne resembles that of a flying pig. No wonder Matt Rayner of the Auckland Museum entitled a diving petrel poster presented at conferences 'Pigs can fly! – unpredicted migration of Common Diving Petrels from New Zealand colonies.' The petrels in question were followed from two colonies, respectively east and west of New Zealand's North Island. During breeding, the parent birds unsurprisingly remained within 300 km of the colony. Once free of responsibility for the chick, they made journeys lasting, on average, eight days and finished 3,000–5,000 km to the south-east in highly productive waters along the Antarctic Polar Front. Thus the reason the petrels chose this region to pass the non-breeding period was far clearer than it was for the Ancient Murrelets.

Far more graceful than the auks and diving petrels is Ross's Gull, whose breast is suffused with pink in the breeding season. This is an almost mythical creature breeding in the high Arctic. There, fewer than 200 nests have been found since the very first scientific specimen was obtained by the supremely handsome James Clark Ross in 1823, not far from the North Magnetic Pole. Even more enigmatic has been the species' wintering grounds. At last, progress in discovering those grounds has been made thanks to an intrepid couple, Mark Maftei and Shanti Davis, and collaborator Mark Mallory.[8] Using data from geolocators and satellite tags deployed at two colonies in the far north of Canada (approx. 75°N), Mark Maftei reported on the gulls' movements at the 2015 World Seabird Conference. Unusually, possibly even unprecedentedly for a scientific presentation, his talk was partly delivered in impeccable rap.

> They're so rare, they're pink, just like a fine steak.
> And you might think you know them but you might be mistaken,
> Cos the fact of the matter is that no-one really knows,
> Where the rarest little gull in the Arctic really goes.
> Their breeding grounds are poorly known and poorly surveyed,
> But as far as they are concerned that's the way it oughta stay
> Cos it's invitation only when they are trying to copulate.

Not only did this enthral the audience, it probably also confirmed that studying Ross's Gulls on a forlorn chill Arctic island involves long periods sheltering in a tent with nothing much to do beyond compose rap! However, impatience was building; the audience wanted to know where the gulls wintered.

> Well they're highly pelagic: they stay far offshore,
> In the icy cold waters off the coast of Labrador.

So we learnt that the birds from the Canadian study colony of Nasaruvaalik concentrated in winter off Labrador in cold seas around 63°N, barely south of the Arctic Circle. Since these seas, on the western side of the Davis Strait, are little visited in winter, it is no wonder that this Ross's Gull locality had never been detected by traditional observation. The wintering grounds of the Ross's Gulls of Siberia are yet to be identified precisely.

These studies show how modern technology is helping fill the gaps in our knowledge of seabird migration, and sometimes provides truly

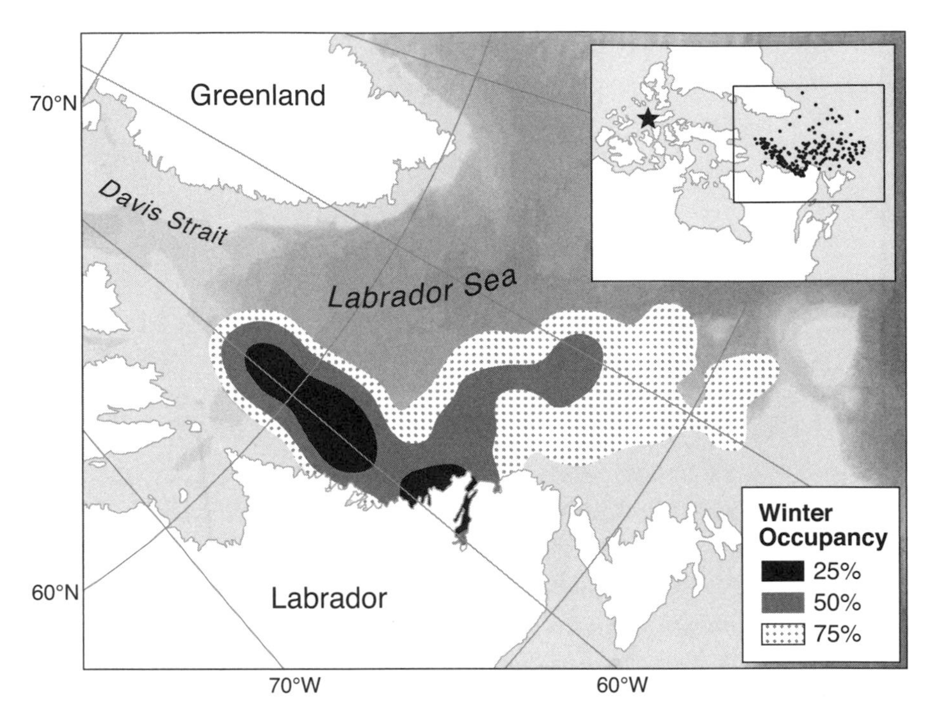


Wintering ranges of three tagged Ross's Gulls in the Labrador Sea from 2011 to 2013 showing 25, 50 and 75% occupancy contours. Inset map shows all winter locations and the black star in the inset shows the breeding site of the three birds on Nasaruvaalik. Map reproduced with permission of Wiley, from the work cited in Note 8, Chapter 4.

startling information. Without question one of the biggest surprises involved Red-necked Phalaropes. These charming small waders spend the summer on northern tarns where, in a reversal of typical roles, the brighter females court a drabber male, lay a clutch, and then devolve all incubating duties onto him. Come winter and the phalaropes qualify as seabirds, spending their days in small flocks picking titbits from the surface. It had always been assumed that the small number of British phalaropes joined greater numbers of their kind from northern Russia to spend the winter in the seas south of Arabia, a known stronghold. By 2012, geolocators had become small enough to be attached safely to phalaropes, of which ten were duly tagged on the Shetland island of Fetlar. One was spotted back on Fetlar the following summer.

Birdwatchers were amazed when it was discovered that a tracked Red-necked Phalarope, breeding in the Shetland Isles, had spent the winter in the equatorial Pacific.

Malcie Smith, the Royal Society for the Protection of Birds' Fetlar warden takes up the story.[9] "We knew through experience that using a walk-in trap was almost always successful with incubating male phalaropes, so I was confident of success and, sure enough, we got our hands on his tag the following morning. We had a well-deserved dram of Scotland's finest that night!

"There was a bit of a problem with having the data interpreted, with colleagues from the RSPB and the Swiss Ornithological Institute becoming involved in making sense of what was pretty messy data. I remember reading emails that included phrases like 'impossible to interpret' which was not encouraging. I was eventually given the 'cleaned up' details by email which nearly knocked me off my seat."

The cause of Smith's unseating was a track that took the bird west across the Atlantic from Shetland to Newfoundland. It then meandered along the eastern seaboard of North America, until crossing central America in mid-September. The next six winter months were spent in the eastern Pacific close to the Equator between continental Ecuador*

* Appropriately the country's name means Equator in Spanish.

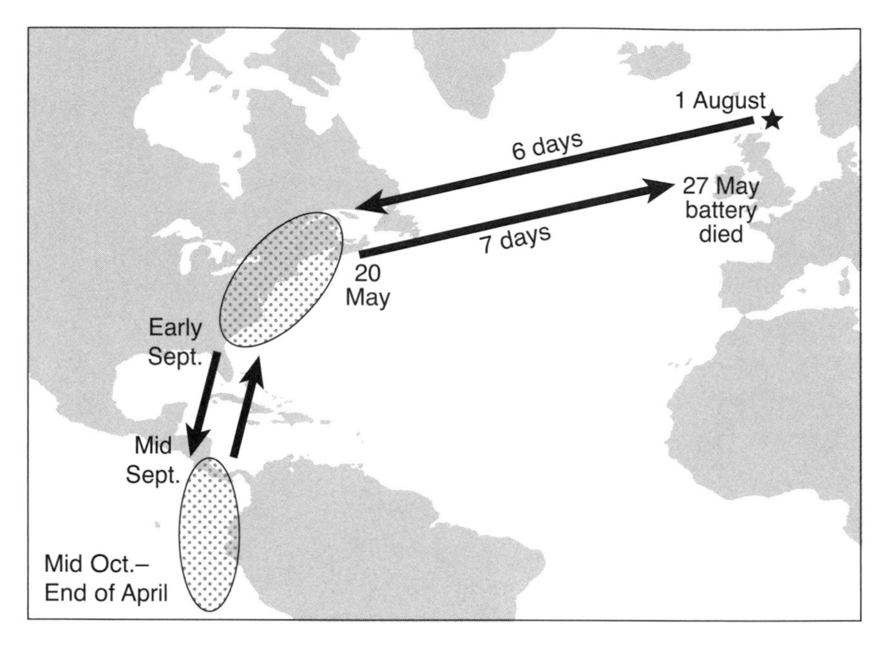


The approximate route and timing of the autumn and spring migrations of a male Red-necked Phalarope breeding in the Shetland Isles and wintering in the Pacific Ocean. Map re-drawn, with permission, from www.bou.org.uk/red-necked-phalarope-pacific-ocean/.

and the Galápagos Islands. Returning to Fetlar in spring, the phalarope more or less recapitulated its southbound route.

Since that study, further work has confirmed that phalaropes from Iceland and Greenland also winter in the Pacific while their fellow phalaropes breeding in northern Scandinavia do head to the Arabian Sea. Without question, this migratory split in the north-east Atlantic was wholly unanticipated.

That said, it is no surprise that different populations of seabirds frequently use different non-breeding areas, reached by different routes. Sooty Shearwaters breeding on islands off New Zealand head to the North Pacific when not breeding, while their Falklands counterparts head to the North Atlantic. So far, so obvious. But as modern technology allows more detailed delving into the variety of patterns, it has become evident that there is an almost infinite variety in the extent to which birds from different colonies, on the one hand, mingle outside the breeding season or, on the other hand, utilise distinct regions.

Minglers include Thin-billed Prions, relatively small grey-and-white petrels whose specialised bill serves to sieve a diet dominated by plankton from the ocean surface. The two principal breeding populations of the prions nest on the Falklands (60°W) in the South Atlantic and, at a similar latitude but one-third of the way around the world to the east, in the French sub-Antarctic archipelago of Iles Kerguelen (70°E). These two populations are both large, perhaps two million and one million pairs respectively, so food in the breeding season is evidently abundant. Yet geolocators attached to birds from the two places showed that the prions that breed 8,000 km apart come together after the breeding season at the Antarctic Polar Front close to the Greenwich meridian (0°W) about halfway between the two colonies. After moulting in this region, the Falklands prions head homewards well ahead of the Kerguelen birds.[10]

Another species where birds from distant breeding populations may encounter one another outside the breeding season is the Ivory Gull *Pagophila eburnea*. The clue is in the names. In plumage, it is as white as ivory, in habits it is pagophilic, which is to say ice-loving. Where there is sea ice, the Ivory Gull can make a living from fish and crustacea, supplemented by more questionable fare like Polar Bear faeces and seal placentae.

These birds have now been satellite tracked from several high Arctic colonies, in Canada,[11] north-east Greenland, Svalbard, and Franz Josef Land east of Svalbard.[12] Although the studies encountered predictable problems arising from the fact that solar-powered satellite transmitters fade into silence aboard an Ivory Gull living in the continuous darkness of the Arctic winter, a fair picture emerged. By mid-winter, Canadian and Greenland birds were both at the ice edge in the southern Davis Strait bordering on the Labrador Sea, while other Greenland gulls, plus those from Svalbard and Franz Josef Land, were all found in south-east Greenland.

Slightly further south in the Atlantic, Danish researcher Morten Frederiksen co-ordinated a geolocator study of Black-legged Kittiwakes on a near-industrial scale.[13] Thanks to the joint efforts of 31 researchers, 439 devices were deployed at colonies spread far and wide; above the

Arctic Circle in Canada, Greenland, Svalbard, and Norway, as well as further south in Iceland, Faeroes, the North Sea, and the western United Kingdom. This spans almost the entirety of the species' North Atlantic range. No fewer than 236 devices were retrieved and provided data. The picture was very clear. Kittiwakes from far and wide generally chose to spend the winter off Newfoundland with an estimated 80% of the 4.5 million adult Kittiwakes in the Atlantic wintering west of the Mid-Atlantic Ridge. Only birds from Ireland and western Britain stayed mainly east of the Ridge towards the European side.

While there was clearly much winter mixing of Kittiwakes from different breeding populations, there was a measure of segregation. I suspect this may be a common pattern. Aevar Petersen, formerly of the Icelandic Museum of Natural History, found that Icelandic Puffins largely overwinter west of Iceland but, broadly, the birds from colonies further north in Iceland spend the winter further north than those from southern colonies.

If birds from separate colonies demonstrably remain apart outside the breeding season, the chance of genetic differentiation and, maybe, eventual speciation increases. An exquisite example comes from the Antipodes. Ignoring a tiny cluster on Great Barrier Island, the world population of Cook's Petrels today nests on two islands at the north and south ends of New Zealand, on Little Barrier Island in the Hauraki Gulf off Auckland and on Codfish Island off Stewart Island respectively. As is true of several other petrel species nesting around New Zealand, Cook's Petrels make a west-to-east migration after breeding, crossing the Pacific to spend the non-breeding period off the western seaboard of the Americas. The story developed when Matt Rayner, a New Zealand ornithologist with a tattoo count that is normal for a building site and above average for a scientific conference, obtained geolocator data from 11 petrels on each of the two principal stations.[14] The southern Codfish birds spent about six months (April–October) in Peruvian waters while the northern Little Barrier birds flew initially to seas off Baja California. Later they headed north-west to the central Pacific before returning to New Zealand.

What makes the work so fascinating is that the geolocator tracking was combined with an investigation of stable isotope signatures of the two groups of Cook's Petrel. These signatures were different and, in both cases, matched the signatures of 100-year-old Cook's Petrel skins residing in the drawers of the California Academy of Sciences and the American Museum of Natural History, and collected either off Baja California or off Peru. Rayner concluded that the very different migratory tracks of the two Cook's Petrel populations have persisted for at least 100 years, probably longer, and this is now reflected in clear genetic differences.

While birds from different colonies may mingle or remain apart outside the breeding season, that, alas, doesn't exhaust the possibilities. Birds from a single colony could potentially use distinct non-breeding areas – and the case of the male and female of a single Sabine's Gull pair heading south from Canada to spend the winter far apart in the Pacific and South Atlantic was mentioned at the start of the chapter.

Consider South Polar Skuas, aggressive attendants at the seabird throngs of the far south, ever ready to snatch an unguarded petrel egg or an unattended penguin chick. The skuas quit Antarctica when winter approaches and, roughly, head north from their breeding area. But on that northward journey, they may encounter a metaphorical fork in the road, a decision point. Geolocator studies of the skuas of Terre Adélie, the section of the Antarctic continent south of Australia, have identified the birds' alternative decisions. They either veered east (right) along the east of Australia to spend the southern winter off Japan, or they veered west (left) to enter the northern Indian Ocean.[15] Just the same parting of the migratory ways was discovered among skuas nesting on the Antarctic Peninsula. There the skuas' decision is either to pass east of Cape Horn into the Atlantic and travel north to seas off Newfoundland, or to veer left into the Pacific where the main non-breeding area is the span of ocean between Hawaii and British Columbia. (See Map 13.) They did both, but individual birds, tracked over 2–3 consecutive years, consistently followed one or other route.[16]

Such individual consistency has emerged repeatedly in recent seabird migration studies. It is a topic to which I shall return in Chapter 7. But while the skuas provide examples of different birds from a single population adopting different migratory strategies, there is another species where individuals from a single population not only head to very different destinations but sometimes shift between different destinations

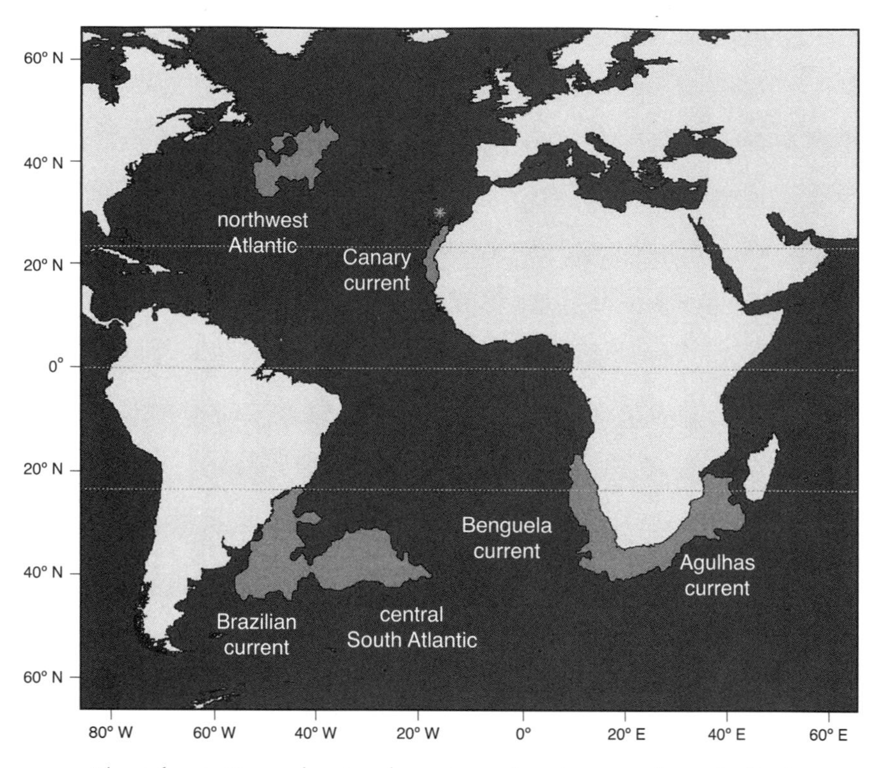


The Atlantic Ocean showing the various wintering areas (grey shading) occupied by Cory's Shearwaters breeding on the Salvage Islands (star). Reproduced with permission of The Royal Society, from the work cited in Note 18, Chapter 4.

in successive winters. That species is Cory's Shearwater, whose nesting stations in the North Atlantic include the Azores and the Canaries. From both, some shearwaters spend the non-breeding (northern) winter period off southern Brazil, whilst others head similarly far south and winter off South Africa, either in the Benguela Current west of the country or 'round the corner' east of the Cape of Good Hope in the Agulhas Current. And, finally, a minority of these shearwaters do not even cross the Equator; they winter in the Canaries Current.[17] What links these very separate wintering areas is that all are characterised by nutrient-rich water upwelling from depth to foster planktonic growth which, eventually, translates into shearwater food such as fish and crustacea.

Now the twist. Maria Dias' team tracked Cory's Shearwaters from the colony on the Salvage Islands, just north of the Canaries. Successive winter journeys were documented for 14 birds, nine of which used the same wintering area in successive winters, mostly the Benguela Current. The other five switched between years, two birds South to North Atlantic, two from the western to eastern South Atlantic and one from the Benguela to Agulhas Currents.[18] Although the switching between years meant birds might spend a winter 7000 km from where they passed the previous winter, the team failed to identify any switching triggers. For example, males were as liable to switch as females, and the birds switching (which were on average in their late teens) were of similar age to those remaining faithful to a wintering area. I wonder whether the various areas may generally offer equally favourable conditions for Cory's Shearwaters, such that visiting one area rather than another has few consequences. If that is right, the average bird might routinely migrate to one area, but occasionally visit others to ensure it was not missing a feeding bonanza. This idea assumes the birds are familiar with all the wintering areas and Maria Dias describes the amazing journeys of one bird that, in two successive winters while 4 and 5 years old and therefore still immature, visited all the wintering areas. To do so entailed flying 108,000 km: "A truly epic feat of flying. I hope it paid off and the bird could make good use of its knowledge of the various areas in later life," says Dias.

* * *

The world, or at least the 70 percent of our world that is sea, has been the canvas as we have explored some of the breathtaking migratory journeys undertaken by seabirds as they travel between their breeding grounds and the sea areas used outside the breeding season. Still to be answered are questions about the details of these journeys. At what speed are migrations undertaken? Are the journeys non-stop or interrupted by periods of rest and recuperation? Given that most seabirds remain with the same breeding partner year after year, do they also manage to remain together outside the breeding season? It is a pleasingly romantic thought. If they do not, are there any differences in the sea areas utilised between males and females, as already outlined for Wandering Albatrosses? And how does the birds' behaviour in the non-breeding season differ from that during breeding?

Mention has already been made of the speed of documented journeys, for instance 300–500 km/day by Arctic Terns, up to 950 km/day by circumnavigating Grey-headed Albatrosses. At first gasp these sustained speeds are stupendous. At a second and more considered glance, they might be anticipated. Depending on species, a seabird in still air probably flies at around 30–60 km/h. Multiply that speed by 20 flying hours, allow the bird four hours of rest, and it is easy to generate a day's track of 500–1,000 km, even without the benefit of a following wind.

Exactly that sort of schedule has emerged from geolocator work I undertook with Tommy Clay of the University of Cambridge. We studied the Murphy's Petrels of Henderson Island in the remotest blue of the South Pacific (24°S). The petrels from Henderson spend the non-breeding period in the central North Pacific around 40–45°N. Twice a year they migrate between the two areas, each time taking only some 10 days to complete the 8,000 km journey.

Now the cunning bit. Geolocators are routinely attached to seabirds' legs, much like a traditional metal ring that can provide survival and movement information if the ringed bird is later found, alive or dead. But, in addition to collecting information on light levels that translates into data on the bird's position, the tiny device can also record whether it is wet or dry. Potentially this allows the researcher to assess as often as every three seconds whether the bird is on or under the sea (device wet), or whether the bird is flying or standing with its feet dry. Using these immersion data, Clay, the analytical force behind our project and a man who confirms that a twinkling eye and statistical expertise are by no means incompatible, worked out that the migrating petrels spent about 75% of the daylight hours and 80% of darkness on the wing. If the petrels' flight speed is around 40 km/h, something of a guess, travelling the necessary 800 km/day is well within their grasp. Our data yielded no hint of any stopovers en route.

While the Murphy's Petrel migration follows a north-south axis, equally rapid journeys are the norm for at least one species following a west-east route. That species is the Westland Petrel, one of the largest burrowing petrels. It nests in the wet dense forest of the west coast of New Zealand's South Island. Once breeding concludes in November, the petrels head east across the Pacific to similar latitudes off Chile and, in the Atlantic, over the Patagonian shelf off Argentina. The 7,000 km journey averages 6 days, and the return in April is barely slower, 10 days. During these journeys, Todd Landers found petrels spent, respectively, a mere 10 and 17 percent of their time on the water.[19]

In general these mammoth seabird treks appear to be undertaken without the prior fattening often observed among landbirds.* This is presumably aided by the fact that seabirds have the possibility of snacking as they travel, if they encounter prey.

In fact, setting aside the species for which flying is impossible, namely penguins,[20†] or enormously hard work, for example auks, the matter could be viewed from an alternate perspective. We might expect the overall speed of long-distance flying seabird journeys to exceed 500 km/day unless interrupted by significant stopovers. Just such stopovers were discovered for the Greenland Arctic Terns mentioned earlier in the chapter, and they have emerged in other studies. Indeed the regularity with which migrating seabirds undertake mid-ocean pit-stops has surprised the research community, especially as the areas used for such stops are evidently favourable and repeatedly utilised. However the areas used vary between different species with different diets and feeding styles.

After breeding in Alaska, Arctic Terns, studied by a group led from the US Fish and Wildlife Service, migrated through the eastern Pacific.[21] Southbound, they utilised a succession of stopover sites, almost as if the road trip had been planned in advance and the motels booked, each chosen for its capacity to provide copious food. The first staging point was off Oregon and northern California where birds lingered for

* The fat Alaskan Bar-tailed Godwits destined for New Zealand were mentioned earlier in the chapter. Many small perching birds also double their body mass before embarking on a major migration, for example a Saharan crossing. Exceptions include the swallows and similar species that can potentially feed on flying insects during their travels.

† The daily travel distance of juvenile King Penguins of 45 km was mentioned in Chapter 2. It is on a par with the 23 km/day achieved by Magellanic Penguins migrating north along the east coast of Argentina.

2–4 weeks. From there, they continued to the next point, off Ecuador and northern Peru. Another stopover followed, off the coast of central Chile. At this point, about 40°S, the Terns turned east and headed across Chile and Argentinian Patagonia. This route of course entailed a traverse of the Andes that took the terns, familiarly known as sea swallows, at least 1,500 m above sea level – but it did avoid the tempestuous seas of Cape Horn.[22] Once reunited with salt water off Argentina, the terns enjoyed a final stop-over on the Patagonian Shelf before proceeding to the Weddell Sea where they encountered their cousins from Greenland!

Most simply, stopover points on migration can be detected from geolocators because the bird remains in a restricted geographical area for several days or weeks. The immersion data allow a more sophisticated, if oblique, approach to identifying stopovers in more aerial species, for example terns and shearwaters. The legs of an actively-migrating flying bird will remain dry. A bird resting in mid-ocean is likely to have wet feet. A feeding bird may quickly alternate between quite short wet and dry spells.

Tim Guilford of the OxNav group is a pioneer of this approach. He studied Manx Shearwaters breeding off the Pembrokeshire coast, west Wales, thereby ensuring that I was fascinated by his study since I had studied these same shearwaters for my doctorate. The migratory route that Guilford reported broadly matched that which had been deduced from ring recoveries. In autumn, the shearwaters first head south as far as west Africa. Then they veer south-west and, aided by the north-east trade winds, approach the coast of Brazil. Onward they fly, barely resting, until reaching the wintering area, rich seas off Argentina about 40°S, comfortably south of Buenos Aires. The northward journey through the North Atlantic back to Wales sees the shearwaters taking a more western route, swinging towards the eastern seaboard of North America and finally approaching Wales from the west. (See Map 4.)

In two respects, the study broke new ground. Firstly, the shearwaters appeared to be spending the winter about 10–15 degrees of latitude further south than ringing recoveries, mostly obtained several decades earlier, had indicated.[23] It is very difficult to know whether this reflects a change in the shearwaters' migratory habits, or whether the ringing

Modern devices have recorded migrations of Long-tailed Skuas between breeding grounds of the High Arctic, north-east Greenland and Svalbard, and wintering regions concentrated along the south-west coast of Africa.

data were biased because a ring recovery is so much more likely on a populated Brazilian beach – think Copacabana – than on a dusty Patagonian shore. Secondly, the work identified a feeding area in the Gulf Stream routinely used by shearwaters heading back to Wales.[24] This area, about 1,000 km east of the coast of the Carolinas, had never been suspected, let alone identified by traditional ringing.

Unusually, and in contrast to the Manx Shearwaters and indeed several other species whose sigmoid track around the Atlantic makes full use of following winds, Sabine's Gulls adopt a relatively straight route both southbound and northbound. When these elegant gulls head south from their breeding station in north-east Greenland, they initially head south before turning left to cross the Atlantic.[25] The gulls skirt but appear not to linger off the Grand Banks, the staging area used by Arctic Terns and also Long-tailed Skuas[26] breeding in the same region of Greenland. Instead Sabine's Gulls press onward to a stopover area in the Bay of Biscay and off Portugal. There they tarry for an average of 45 days before continuing south to spend the winter in the Benguela Current. On the return migration, the initial phase is rapid; 6,000 km covered at 800 km/day takes the birds to a stopover region off Morocco, Mauritania, and Senegal. The time spent thereabouts is 19 days, much less than is spent in the different Biscayan stopover region in autumn. Then, from west Africa, the birds fly fairly directly to Greenland to round off a 30,000 km round trip – and that impressive total excludes local meandering in the Benguela region. Why Sabine's Gulls eschew the chance to benefit from following winds, a tactic used by other species, is not obvious. How seabirds often exploit prevailing wind patterns to minimize the costs of their travels is the subject of Chapter 6.

* * *

Seabirds are paragons of fidelity. Normally, the two members of a pair will continue to breed together until death do them part. Numerous studies have shown that this fidelity is rewarded by increased breeding output. Conversely, divorce is costly, not least because it may take several years until another partner is found, and those are wasted, chickless years.

It is not unknown for bird pairs to remain together on migration. Geese and swans achieve this romantic trick, and remain together with their youngsters on the wintering grounds, having perhaps migrated *en famille* from Siberia. This enables researchers to assess the breeding success of known pairs without ever needing to venture to Siberia.

The temptation therefore to wonder whether seabirds might also remain together on migration is strong. But, oh, the difficulties. Think of a thousand gannets arrowing into the water to feast on a mackerel shoal. While the male gannet finds himself sitting on the water, trying to swallow a large, lively and not-to-be relinquished fish, his mate wanders away with the crowd because she has yet to catch a fish. Imagine the hours of darkness when the breaking waves of a stormy sea and the keening of the cold night wind render maintaining contact by sight or sound overwhelmingly difficult. It seems improbable that pairs could maintain contact; yet the ability of young auks to remain with their parents over periods of weeks argues that the improbable might turn out occasionally to be true. Alas, it is a truth yet to be demonstrated.

Species such as Wandering Albatrosses where immature and adult males and females utilise different sea areas are clearly not candidates. But how about a species where, as far as is known, male and female migrate to similar regions? Scopoli's Shearwater is one such, and work at a large Italian colony addressed the question of togetherness outside the breeding season[27] when the shearwaters leave the Mediterranean and head south along the western margin of the African continent, some going no further than Mauretania, others reaching as far south as Namibia. Members of a pair tend to spend a like number of days travelling and so reach similar wintering areas. But the geolocator data provide no hint that they were journeying together. How they come to use similar – but not identical – wintering areas remains unknown. Conceivably the colony is structured such that genetically similar birds, with a tendency to undertake similar migrations, breed in the same part of the colony and sometimes with each other.

Let us turn the question on its head. If the evidence that pairs of seabirds travel together is, at best, inconclusive, are there species where males and females are likely apart when not breeding? The separation of female Wandering Albatrosses to the north and males to the south could be related to structural differences between the sexes. Because males are heavier, the extent of wing area available to support each kilogram of bird is lower; in the jargon, the wing-loading is greater. Males might then be able more efficiently to exploit zones of the Southern Ocean where the winds are stronger, namely the Furious Fifties to the south of the (merely) Roaring Forties.

Another reason for males and females to have different migrations might be their contrasting roles early in the breeding season. It normally falls to the male to secure a nesting site, maybe a burrow in a dense colony. To prevent usurpation of that site, it could pay to arrive early. Exactly this argument is used for migrant landbirds where, in many species, males winter north of females (in the northern hemisphere), and arrive on the breeding grounds before females.

Those Cory's Shearwaters nesting on the Salvages mostly head to the southern hemisphere after breeding. When they return in spring, the journey starts slowly, around 400 km/day with about half the 24 hours spent in flight. But, north of the Equator, the pace picks up to around 1,000 km/day, with three-quarters of the day in flight. Digesting the data in Maria Dias's paper,[28] I found it difficult to resist the temptation to mumble *sotto voce* that the birds were anxious to get home as soon as possible.

However a small minority, 8 percent, remain in the Canary Current near the Salvages colony.[29] These 'stay-at-home' shearwaters are overwhelmingly males and, on average, they arrive back at the colony three weeks earlier than the birds that had visited the southern hemisphere. It is the late-arriving birds that may have to skip breeding because there is no longer room at the inn and all breeding sites are already tenanted. There is therefore an apparently good reason for male Cory's Shearwaters to winter north of females.

Unfortunately this tidy explanation of a gender difference in the choice of non-breeding area appears to founder for another species where males and females have somewhat different migratory destinations. The Balearic Shearwater is a Critically Endangered species breeding in the western Mediterranean. After breeding, the birds pass through the Strait of Gibraltar in early July and spend three months in the eastern Atlantic, favouring two areas, one off Portugal and the other off Brittany. From an overall sample of 26, a study led by Tim Guilford tantalisingly reported that the five birds choosing Brittany waters were all females.[30] Subsequent work by Rhiannon Meir of Southampton's Institute of Oceanography has confirmed that males remain off Iberia whilst females occur in roughly equal numbers off Iberia and Brittany. However, from late September the birds, now clad in fresh plumage, have returned to the Mediterranean and indeed visit the colony sporadically for five months before their late February laying. It is difficult to conceive that the apparently different distribution of males and females during the Atlantic phase of the year has any impact on burrow occupancy several months later.

* * *

The birds have finished breeding, and some have migrated immense distances. Several leisurely months lie ahead before the birds are compelled to return to the breeding colony. Or perhaps the notion of leisure is fanciful if the bird is a Northern Fulmar competing with 500 other fulmars behind a trawler, or a gull with hours to wait before it can join the gull melee to open the next batch of garbage bags delivered to the rubbish tip.

In fact there are at least two key factors that may influence how a seabird's daily schedule outside the breeding season might differ from that during breeding. The first is that it is the period when most species moult. They normally do not moult during breeding, and instead grow the next year's brand new set of feathers outside the breeding season. Very likely this is because the moult is energetically demanding. As is true of all penguins, Macaroni and Rockhopper Penguins remain ashore during the moult period. They look thoroughly dejected as the old feathers, pushed out by the growing fresh feathers, flutter over the colony that acquires the look of a down duvet apocalypse zone. The penguins' daily energy expenditure during this time is about 40 percent higher than during incubation.[31]

For species which undertake a long post-breeding migration, moult generally occurs after the journey. The reason is clear. While the principal wing feathers are being shed and re-grown, the seabird's flying ability is potentially reduced. For most species, where only a few feathers are shed at a time, flight may be impaired but not lost. For larger auks, whose small wings relative to body size render flight taxing at the best of times, moult may be a period of flightlessness. Indeed if several wing feathers are moulted at once, the hazardous process may be completed more quickly than if they adopted the more normal sequential strategy of other birds.

The second factor likely to influence seabird activity outside the breeding season is that birds are then freed of the requirement to shuttle back and forth between colony and feeding area. This may be a relatively trivial factor for, say, a gull or cormorant that is feeding close to its colony when breeding, and feeding close to an onshore roosting site in winter. It could be far more significant for a petrel whose breeding season responsibilities include flying several hundred kilometres every day. Indeed the prudent petrel might reach a non-breeding sea area with adequate food and, in the vernacular, 'chill out', flying only as far as required to meet its much-reduced daily energy needs. Presumably this switch to a life of relative inactivity would be even more pronounced if the area chosen was rich in food, be it natural or fishery waste. We might bet on a prudent petrel choosing such an area.

While some penguins are adrift in the vastness of the Southern Ocean when not breeding, that is not true of Gentoo Penguins satellite-tracked from South Georgia. British Antarctic Survey researchers reported that "five penguins tracked from Bird Island [South Georgia] remained close inshore, and although they did not return to the initial tagging site, they did appear to return to land each evening. They made diurnal trips to sea of similar distance from land as those during the breeding season, even though the constraints of chick rearing were absent."[32] It was entirely business as usual.

Perhaps the most striking contrast between breeding season and non-breeding season daily routines has emerged from species which, outside the breeding season, sever their link with *terra firma*. Here the combination of moult, possibly impaired flying ability, the need to conserve energy, and the possibility that migration has taken the bird to a rich feeding area would all point towards relative inactivity. Take the case of South Polar Skuas.[33] Migrating northwards and returning southwards, they spend a little less than half the 24 hours in flight – and therefore just over half the entire day sitting on the water. But when they reach their destination, say the Seychelles or the Bay of Bengal, they may spend virtually all night on the water and a mere 15 percent of the daylight hours in flight. Such a low activity life-of-Riley could be the result of impaired flight or because food was easily acquired or both; with the skuas, we do not know which.

A comparable contrast between activity patterns during the breeding and non-breeding seasons has been discovered in several medium-sized petrels, for example White-chinned and Chatham Petrels which remain in the Southern Ocean when not breeding. Tommy Clay and I uncovered just the same contrast in the Murphy's Petrels we studied. When off-duty during incubation and making trips of 10,000–15,000 km around the South Pacific (Chapter 5), they spend around 95% of their time at sea in flight. Their activity levels cannot be faulted. But when Murphy's Petrels are passing the non-breeding season, the (northern) winter, in the North Pacific, 80 percent of darkness and 60 percent of daylight is spent sitting on the water.* In other words, the birds quit the South Pacific to spend their non-breeding period sitting out North Pacific winter storms and enduring 16-hour nights. Natural selection does indeed propel animals along curious routes.

While Clay and I failed to identify when Murphy's Petrels moulted during their North Pacific sojourn, this knowledge gap has been addressed in another recent study. A group from the French Centre National de la Recherche Scientifique carried out a project on three small petrels nesting in the sub-Antarctic Kerguelen archipelago.[34] The three species were Antarctic and Thin-billed Prions and Blue Petrels. All spend the non-breeding period, about 8 months, in the Southern Ocean. During that time they moult, a process that lasts two and a half months in Thin-billed Prions and Blue Petrels, a month longer in Antarctic Prions. But how does the researcher assess when that shedding and regrowing of the feathers is happening in birds weighing barely 200 g fluttering among the white caps of the Roaring Forties?

For each species, the immersion sensors on the birds' legs show a distinct period, corresponding to the moult, when the daily time with wet feet climbs from two hours to almost ten. That is when the main flight feathers moult. Unexpectedly this moulting period occurs in the early part of the non-breeding period for Thin-billed Prions and Blue Petrels and at the end for Antarctic Prions. As sketched earlier in the chapter, the Thin-billed Prions have headed west from Iles Kerguelen to an area south of the Cape of Good Hope and also south of the Antarctic Polar Front. This is where they moult, as do the Blue Petrels. In stark contrast – and this was confirmed by the stable isotope picture (Carbon-13) – the Antarctic Prions moulted about 15 degrees of latitude further north, many of them east of Kerguelen towards Australia and New Zealand. Besides pinpointing when and where the moult takes place, a task completely beyond the capacity of seabird biologists 20 years ago, the study raises many questions. For example, why do such species as Thin-billed and Antarctic Prions, similar species that are frustratingly difficult to tell apart, organize their years in such different ways?

* The interaction between flying, sitting on the water, and feeding will explored in Chapter 9.

While activity patterns indicate and sometimes pinpoint where and when moult occurs, the most dramatic signal would come from species that lose the power of flight altogether during moult. This is what larger auks do, but they routinely spend most of their non-breeding lives on or under the water so the change in daily routine is not very marked. Ecologically equivalent to the auks are the diving petrels of the Southern Hemisphere. The diving petrels tracked by Matt Rayner to the Antarctic Polar Front spend up to 95 percent of the 24 hours on the water in the middle of their non-breeding period, and are probably nearly flightless.

Rightly renowned as masters of flight, albatrosses replace their principal wing feathers, the primaries, in a complex sequence. Sometimes spread over two years, the process has been interpreted as an adaptation to ensure albatrosses maintain that magisterial status, and never become floating and flightless, unable to feed. This interpretation probably stands for the higher latitude species of the Southern Hemisphere but it falters for lower latitude species. Step forward the Laysan and Black-footed Albatrosses that despatch their chicks from Midway, at the north-western end of the Hawaiian chain, in June or early July. The breeding adults then head into the North Pacific and concentrate off Kamchatka in the case of the Laysans, or slightly further south off Japan and in an alternate area around the Aleutians in the case of the Black-footed Albatrosses. Following this northward journey, between 30 and 70 days after quitting the colony, there is a period of some three weeks when sustained flight bouts disappear from the birds' daily routine. All birds of both species spent at least one full day during this time entirely floating on the water.[35] Further, Laysan Albatrosses on average spent seven days floating on the water's surface for more than 90 percent of the day and 16 days floating for more than 80 percent of the day. The pattern for Black-footed Albatrosses was very similar. Clearly the birds were virtually flightless, only foraging intermittently, a state from which they 'recovered' in the ensuing months.

Today the scarcest North Pacific albatross is the Short-tailed. While formerly the species was so abundant on the Japanese island of Torishima (Izu Islands) that a light railway was needed to transport feathers from the killing fields, the nesting areas, to the beach prior to export, today's global population numbers around 1,500 birds. One of those birds, tracked during its moulting period, spent no fewer than 39 days drifting on the water among the islands of the Aleutian Arc. Daily it drifted to north and to south as the tide ebbed and flowed in these productive waters. Presumably enough food came within reach of the bird's mighty bill to sustain life until flight could be resumed.

So much of the information in this chapter has come from geolocators, miraculous packets of electronics now smaller than a broad bean. Attached to seabirds weighing 100 g and upwards, they have transformed our knowledge of the truly astronomical distances covered by seabirds, plus the routes and the rests taken. When combined with superficially simple information on whether the bird is flying or swimming, it is possible for an extraordinarily detailed picture of a seabird's non-breeding adventures to emerge. Doubtless, in the next few years, the devices will be further miniaturized so that they can be attached to the very smallest species, such as storm petrels. However the relative coarseness of the positional information, often no better than 200 km, means that geolocators fail to provide truly detailed information on a bird's comings and goings during the breeding season. The necessary precision is available from GPS trackers which are key aids in understanding activities of breeding birds.

# A Navigational Diversion

Seabirds routinely make prodigious journeys across immense oceans, feats which instantly raise questions about their powers of navigation. Conceptually, navigation can be divided into two more or less distinct elements. The first is orientation, the capacity to follow a roughly direct route from the present position to the destination. In humans, this might involve following a compass direction from Point A to Point B. In birds, it might involve flying in a straight line between the breeding area and the wintering grounds, guided by an internal compass using cues whose identity I shall shortly address. The second element of successful navigation is a 'map' of the world, a sense of one's present position in relation to a destination.

Remarkably, the first convincing demonstration that birds indeed have a map sense involved a familiar seabird, the Manx Shearwater. The experiments, conducted by the late Geoffrey Matthews in the 1950s, involved transporting breeding birds from their burrows to release points that were hundreds of kilometres from the colony and, crucially, were beyond the areas that even so prodigious a wanderer as the Manx Shearwater would ever have visited.

Arguably the key tests were undertaken on incubating birds, known from prior experiments to be the most ready to make a rapid passage homeward. Transported from the colony on Skokholm, west of Wales, to various inland points around England, the birds showed a strong tendency to head immediately towards Skokholm when released under sunny skies. And some birds reached their home burrow and waiting egg so rapidly that they must have flown virtually straight across country. The tendency to orient towards home at the outset largely vanished if the birds were released under heavy cloud.

Intriguingly some birds were consistently quick at returning, others consistently slow. Such consistency would not be anticipated if a rapid return was simply due to a random, albeit fortunate, choice of route, and so that very consistency among individuals is evidence of navigational ability.

These experiments from yesteryear suggested that the shearwaters possess some map of the world, and a compass that utilises the sun, possibly in conjunction with other cues. Since then, the study of bird navigation has hugely advanced. However most of the new information has derived not from seabirds but from landbirds, not least because the latter are so much easier to keep in captivity, and there subject to experimentation. It is therefore timely to detour briefly into what is known about landbird navigation, before pondering whether the findings can be extended to seabirds.

A sun compass is likely widespread among birds. A common experiment involves shifting the internal clock of a captive bird by altering the light regime within its cage such that 'dawn' and 'dusk' occur, say, six hours later than in the outside world. When the bird is displaced some distance from home, and released on a sunny day, it will, if using a sun compass for navigation, assess the position of the sun. Now comes the problem. If the northern-hemisphere bird is released at a time its internal clock considers midday, and it needs to fly south to reach home, it will fly towards the sun. Alas for the bird, the real time is 6 pm in the late afternoon. Flying towards the sun takes the bird west, a 90 degree error. By means of such clock-shift experiments, the importance of a sun compass to birds has been repeatedly confirmed.

Night falls and the sun dips below the horizon. But, in the absence of cloud, the stars shine bright in positions that differ according to the position of the observer – or bird – on the ground below. Experiments have confirmed the importance of the starry heavens in determining the direction in which migrating birds head.

Many such experiments exploit the fact that, during the migration seasons, small nocturnally-migrating birds are restless at night, typically fluttering in their cage while pointing in the direction they would migrate if unrestrained. This preference can be recorded if the circular cage floor is inked. The concentration of inky footprints in one sector of the floor indicates the birds' preferred direction. In a classic 1970 study of a widespread summer migrant to eastern North America, Steve Emlen of Cornell University hand-raised three groups of Indigo Buntings variously isolated from the stars.[1] Those prevented from viewing the night sky prior to the autumn migration season could not select the normal migration direction, south, when tested under planetarium skies. By contrast, those exposed as juveniles to a normal, rotating, planetarium sky could pitter-patter towards the species' naturally-preferred southerly direction. A third group was exposed to an incorrect planetarium sky in which the stars rotated about a fictitious axis. When tested during the autumn, these birds took up the 'correct' migration direction relative to the stars' new axis of rotation. These results suggest that the Indigo Buntings use the stars to navigate, a process that has to be learnt and is based on the axis of celestial rotation.

When the sky is cloudy, sun and stars may be of little use but the earth's magnetic field remains unaffected. Since the first demonstration of the ability of birds to use this field in the 1960s, the ability has been discovered in some 20 migrant species.[2] A typical test involves a bird active in a cage similar to that described above. When the magnetic field inside the cage is altered from the earth's, the bird's preferred direction alters. If, for example, the field is reversed by 180°, a bird that was fluttering towards the north starts fluttering southwards.

Because birds seem to use smell less in their daily comings and goings than mammals, there has been historical resistance to the idea that olfaction might help birds overcome navigational challenges. Moreover, the landbird experiments, mostly with homing pigeons, have yielded conflicting results. That said, the balance of evidence is that smell is important if pigeons are to reach home successfully.[3] But, as we shall shortly see, the evidence that smell may guide seabirds (as opposed to landbirds) over substantial distances is accumulating.[4]

While the sun, stars, and magnetic field can, at least in theory, provide the bird with positional information, even when it is beyond the boundaries of the places it has ever visited, it is tricky to imagine how that might apply to the olfactory landscape, or 'smell-scape'. Knowledge of that smell-scape can presumably accrue only piecemeal as the adventure of life takes the animal beyond its current boundaries.

The same is true of landmarks. The closer a bird is to home, the more likely it is to be familiar with local landmarks, lessening any reliance on the various compasses just discussed. And the use birds make of such local landmarks has become apparent as GPS-tracking has made it possible to plot the precise route followed by homing birds. For example, different pigeons released repeatedly from the same locality near Oxford, England, developed personally-stereotyped routes to the home loft. These routes were by no means the straightest possible. Rather they appeared to connect a series of landmarks, or landscape features such as main roads, leading in roughly the right direction.[5]

The nuances of how a landbird might integrate information from the sun and stars above, from its sense of the earth's magnetic field, and from its familiarity with the visible and olfactory landscape is beyond the present discussion. It is time to ask whether the same abilities – or others – are extended to seabirds.

From Geoffrey Matthews' pioneer experiments with Manx Shearwaters came hints of the use of the sun. This possibility has recently been given further credence by Ollie Padget of the OxNav group. Instead of clock-shifting caged birds, the approach traditionally applied to landbirds, Padget was able to clock-shift incubating Manx Shearwaters while they sat on the egg in the burrow. When the subject shearwaters were taken some 50 km from the colony on Skomer, and released, their homeward journeys differed from the 'correct' route in a manner consistent with the time shift. Even though the birds were released in seas they will assuredly have visited previously in their lives, the clock-shift effect persisted right until the birds were within 8 km of Skomer, at which distance the breeding island was certainly in view.

Although the use of a magnetic sense by migrating landbirds is beyond dispute, a similar sense has yet to be demonstrated in seabirds. Indeed the failure of several tests specifically designed to deliver evidence of such a sense gives reason to wonder whether seabirds absolutely lack this ability. For example, Wandering and Black-browed Albatrosses and White-chinned Petrels have all been despatched across the Southern Ocean bearing head-mounted magnets sufficiently strong to confound the birds' perception of the earth's magnetic field.[6] No test yielded any evidence that the bird's ability to forage successfully or to return expeditiously to its home colony was impaired. In every experiment, other senses, notably sight and smell, were available to aid the magnetically-impaired birds so the evidence for a lack of magnetic sense is by no means conclusive. But one wonders . . .

Given that smell helps seabirds find food (Chapter 8), it is no surprise that there is good evidence that it also helps them traverse their salty domain. One strand of evidence comes from Scopoli's Shearwaters tested from small colonies off the Tuscan coast of western Italy.[7] At the beginning of an incubation stint,* the subject shearwaters were ferried about 400 km westward to a release point around 100 km south of Marseille and out of sight of land. Released birds fell into three groups; control birds with no manipulation, magnetically disrupted birds carrying a magnet glued to their heads, and birds unable to smell and termed anosmic. The anosmic state was achieved by washing the olfactory mucosa with zinc sulphate. This chemical treatment temporarily knocks out of the sense of smell which returns in a matter of weeks.

Carrying tracking devices that allowed the research to plot the routes taken, control and magnetically disrupted birds returned home fairly directly. Apparently they could assess where they had been released and orient accordingly, towards the east. Not so the anosmic birds. Their initial orientation was towards the north, the French coast. Once in sight of presumably familiar landmarks, they tended to follow the coast homeward, and therefore took longer to reach the colony than the other shearwaters.

Navigating seabirds variously use information from the sun, from landmarks ashore, and from the pattern of smells, precise identity unknown, carpeting the sea. But there remain further possibilities to be explored. It would surprise me if they did not recognize the alteration of the cloud patterns that can signal to mariners the presence of land not yet visible over the horizon. More tantalizing – and totally without any support at the time of writing – is the thought that seabirds, like the Polynesian navigators of a thousand years ago, might be able to recognize wave patterns on an oceanic scale and use these patterns, created by the consistent interplay of prevailing winds and fixed land obstacles, to assess their position.

* Since incubation stints last a week or more, a bird removed from its egg early in the stint has no immediate need to feed. It is also likely to be strongly motivated to return to its nesting burrow. In the experiment any eggs at risk of predation in the parents' absence were artificially incubated.

CHAPTER 5

# Tied to Home

## Adult Movements during the Breeding Season

Literature would be the poorer without the opportunity to chronicle the joy and the problems encountered by people as they reproduce and create the next generation. While it is difficult to assess what joy seabirds reap, they unquestionably face problems when embarking on reproduction.

The first problem is that they must nest onshore, often on cliffs or islands safe from predators. Such places are not necessarily close to the best feeding grounds, especially if the species is one specialising on low-density food scattered across distant oceans. An inevitable consequence is that, however efficient is a seabird's flight, large amounts of time and effort will necessarily be devoted to, or one might say wasted on, travel between breeding colony and feeding area. This is likely to restrict the number of young that can be reared.

The second problem follows from the first. Since eggs or small chicks would die if left unguarded for any significant period whilst an adult was away at sea, a single-parent family is not an option for seabirds. At no stage of the breeding cycle is this more true than during incubation. Once one or several eggs are laid, more or less non-stop parental attention is crucial if the eggs are to hatch. This applies in Antarctica where the threat may arise from fierce blizzards, and it applies equally in a cliff colony of Common Guillemots in Great Britain where gulls are perpetually circling in the hope of grabbing an unattended egg. It remains true in the tropics where the predators might be crabs or frigatebirds.

Parental co-ordination between the two pair members is then a vital element of successful breeding. It obviously starts well before laying. Birds return to the colony. *The* One is sought and found. A bond is formed, or re-formed since seabirds so very often retain the same mate from one breeding attempt to the next. Then copulation occurs and in due course an egg or eggs are laid, and incubation starts.

That overly brief account of the pre-laying period rushes past variation in its pattern between various seabirds groups. For some groups, cormorants and gulls for example, it is a period that unfolds in a month or thereabouts, broadly as described, and minimal extra knowledge has arisen from modern devices. Other groups depart intriguingly from the basic pattern.

Among some auks, the first return to the colony can be earlier than might be anticipated, in September or October, well before the dark depths of the northern winter.[1] This is true of southern populations of the Common Guillemot while more northern populations do not return until spring simply because the breeding cliffs are snow-clad through the winter. As spring draws nearer, so a cyclic pattern develops among several auk species: three days of colony attendance followed by 3–5 days of absence. Very early attendance at the colony perhaps helps the bird secure ownership of a site when competition for a limited supply is fierce, and perhaps helps ensure male and female are physiologically synchronised for the forthcoming breeding effort.

That time is often of the essence is neatly illustrated by the migrant Common Terns returning with implanted PIT tags to Peter Becker's artificial German islands (Chapter 1). The tags yield robust information on the precise date of a bird's return. When a male and a female that had previously bred together both survived the winter at sea and reached the colony in Germany, reuniting was the norm. When reuniting, the two members of pairs typically arrived back within two days of each other. But around one-fifth of pairs that survived divorced, and their return dates typically differed by a week. If the two erstwhile mates arrived 16 or more days apart, they always split. Possibly the time pressure is such that the first-arriving tern starts seeking a mate as soon as s/he arrives at the colony. If the former mate returns within that search window, they reunite, If s/he does not . . . well, too bad. It is better for the first-arriving bird to establish a new partnership than to forgo breeding altogether by waiting for too long.[2]

As breeding approaches, penguins, too, return to their colonies, sleek and plump. Once ashore, the species that have dispersed far outside the breeding season, for example Macaroni* and Rockhopper Penguins, remain ashore until laying. The female has come ashore with sufficient body reserves to form the egg before she heads back to sea immediately after laying. Likewise her mate has sufficient reserves to remain ashore, initially during the pre-laying period, and latterly during his incubation stint, the clutch's first.

Arguably the pattern of pre-laying attendance shows most variation among the petrels and albatrosses. There is a batch of non-migratory species, including the Northern Fulmar, some prions, Kerguelen Petrel and Blue Petrel, which begin to visit the colony once the post-nuptial moult is completed. This may be as little as 2–3 months after the previous chick has fledged. More commonly seen is a progressive increase in numbers attending the colony in the months prior to breeding, with males often arriving a little earlier than females. However the peak in numbers is not seen immediately before laying. Rather, it occurs significantly

* As I wrote that, I was aware that extravagantly-plumed Macaroni Penguins were named after macaronis, extravagantly-dressed eighteenth century dandies. Then the question arose of what might be the connection between a dandy and those familiar tubes of pasta. The answer seems to be that the founders of the Macaroni Club wished to celebrate their exotic fashion sense by alluding to a food that was exotically novel in England 250 years ago, namely macaroni.

before laying, and this interval between peak attendance and laying varies surprisingly between species, from a maximum of 2–3 months in the Grey-faced Petrel to a minimum of some 10 days in the South Georgia Diving Petrel.

When numbers peak, mating occurs, leaving the female, duly inseminated, free to head to sea to feed up in preparation for laying. In certain species, for example Manx and Balearic Shearwaters, the male continues to visit the colony by night, presumably to 'keep an eye' on the burrow, while spending his days at sea. Meanwhile the females are absent from the colony for around two weeks before returning one night to lay.

This pre-laying absence from the colony or exodus is also known as the honeymoon period. It is a wonderfully inappropriate name since the sexes are very possibly apart, and the female, already inseminated, is carrying the sperm whose exchange is viewed as a key part of a happy honeymoon. Not surprisingly where the behaviour of the two sexes is very different, the distribution at sea is also different. Tim Guilford found that, during this period, male Manx Shearwaters remained within some 300 km of the Pembrokeshire study colony on Skomer, a distance compatible with their return to land on most nights. Meanwhile the females went further afield to the south-west, beyond the continental shelf and into the Bay of Biscay, over 1,000 km from the colony.[3]

Most albatrosses show a similar pattern to the Manx Shearwater, the males lingering moderately near the colony while the females head away on the exodus. It can take them on impressive journeys. For example, Atlantic Yellow-nosed Albatrosses breeding on the mid-Atlantic island of Gough may head 2,500–3,000 km to the Benguela Current off South Africa.[4]

An intermediate pattern is seen in species where the male does quit the colony during pre-laying but for a lesser period than the females. Enter the Northern Fulmars breeding on Eynhallow, in the Orkneys, and long studied by Aberdeen University researchers. Over excursions lasting up to 32 days, birds reached a maximum of 2,900 km from the colony. Males (18 days) spent less time away than females (25 days), and the majority of males remained within the North Sea. On the other hand most females flew north towards the Norwegian/Barents Sea, some reaching North Cape at the extreme north of Norway.

Finally there are species, for example the Short-tailed Shearwater, where the males also depart the colony at about the same time as females to prepare for the first long incubation spell. There is a mass exodus and the colony is virtually empty in the run-up to laying.

Perhaps unexpected is the recent evidence from tracking studies that males and females may head to distinctly different sea areas, even when both are absent from the colony for a lengthy and roughly similar period. Murphy's Petrels, studied on Henderson Island by Tommy Clay and myself, have a pre-laying exodus that lasts about six weeks. Birds head south-west but the average maximum distance from the colony is 900 km greater for males (3,800 km) than females (2,900 km). Further east in the Pacific, Matt Rayner discovered a comparable difference among Chatham Petrels, the males' exodus taking them, on average, 1,500 km further from the colony on Rangatira Island* than the females' trek.[5] However, perhaps the most striking example comes from Barau's Petrels.

Barau's Petrels were first described as recently as 1963. Most of the world population nests near the jagged volcanic peaks of the French Indian Ocean island of La Réunion. Soaring up to 3,000 metres, the peaks generate swirling updrafts in the late afternoon which the petrels exploit as elevators. When the petrels are lifted 2,000–3,000 m upwards from sea level to their high-altitude colonies as dusk approaches, so the returning birds are saved energy. I observed this behaviour in 1974 when sharing a mountain-top cave with local Creole peasants who were enjoying a night out in the wilds. Not only was the cave shared. So too was the chicken, which arrived alive and became chicken curry after a penknife sawed open its throat.

For 6–7 weeks before laying from late September to early November, the Barau's Petrels embark on their exodus. In the case of females, it takes them 1,500–2,000 km from La Réunion to seas south-east of southern Madagascar.[6] Males generally go 1,000 km further to the waters of the Agulhas slope off Mozambique and north-eastern South Africa. Associated with this geographic difference is a difference in behaviour. The overall trip speed is 21 km/h for males and a lower 15 km/h for females,

* The study island in the Chatham group, east of New Zealand.

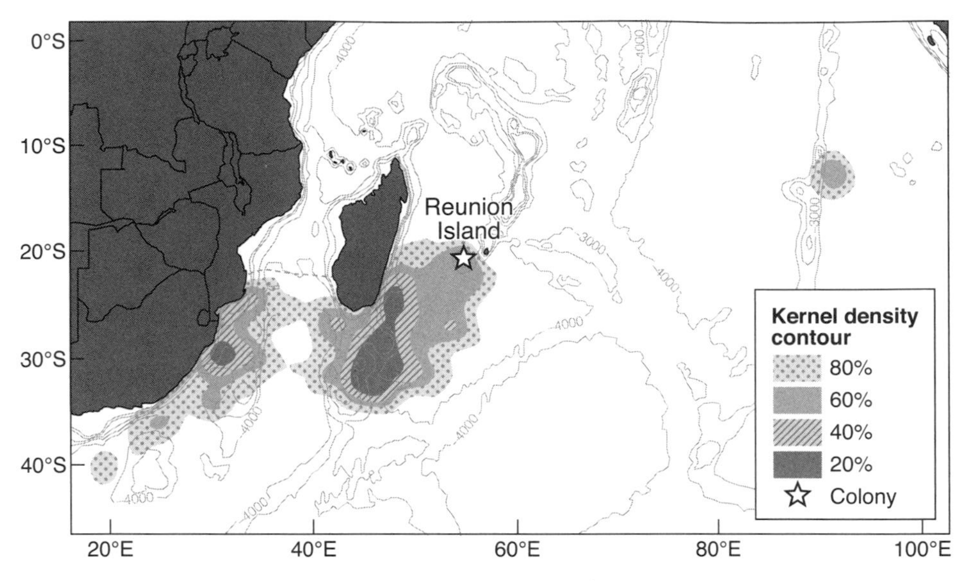


Two distinct areas are used by Barau's Petrels during their pre-laying exodus from the Reunion colony, one off the African coast, the other south-east of Madagascar. The former is mostly used by males, the latter by females. Shading intensity corresponds to intensity of use by the birds. Depth contours (metres) show the birds are mostly using sea areas where the depth is greater than a kilometre. Reproduced with permission from Elsevier, from the work cited in Note 6, Chapter 5.

which partly arises because females spend more time sitting on the water in darkness. Why there should be this consistent difference among gadfly petrels, with males travelling further than females during the pre-laying exodus, remains something of a mystery. Perhaps it benefits females to minimize the distance that needs to be flown when they are laden with a nearly fully-formed egg.

* * *

With eggs to incubate and then chicks to tend, parent seabirds are significantly tied to one spot, the colony. They need to return there more or less regularly, otherwise the breeding attempt will founder. This constraint, this leash to home, inevitably restricts the journeys that can be made during the breeding season. It also makes the case for positioning colonies closer to, rather than further from, prime feeding areas - if this is feasible. Philip Ashmole is a seabird biologist who has thought and written influentially about these issues.

It is satisfying to have one's name commemorated in the name of an animal, even if it is a mere bug. It must be even more satisfying to be linked to an important idea that influences the trajectory of scientific thought and research. But, oh, the saintly joy of being linked to a halo! That beatific joy has befallen Ashmole. As a member of the British Ornithologists' Union Centenary Expedition to Ascension Island in the early 1960s, he studied seabirds under the tropical Atlantic sun. His thoughts turned to how the numbers of tropical oceanic species might be regulated.[7] Predation and disease were having little obvious impact, and space for extra breeding pairs remained available. Outside the breeding season immense numbers of birds could spread over such an expanse of ocean that food might not then be in short supply. But the situation was possibly different when the birds were assembled at the colony while breeding. Then, as numbers at the breeding site grew, so the birds would deplete food in an expanding 'halo' surrounding the colony, obliging them to travel further from the colony and/or spend more time feeding. Eventually this would result in reduced breeding success. Growth of the birds' population would halt, especially if the increasing size of the colony led to ever-longer deferment of the start of breeding.

Ashmole's halo is a useful prism through which to observe many of the results that have emerged as birds' forays from colonies during incubation and chick-rearing have been documented, particularly by GPS tracking. Not only will I investigate the extent of those forays but also how they are affected by the size of the colony and the proximity of neighbouring colonies.

Once the clutch is laid, parental co-ordination is a vital element of successful incubation. And the stakes are high. Seabird eggs tend to be large for the size of bird, and represent a substantial physiological investment. Nowhere is this more relevant than in the smaller species that lay but a single egg, for example the storm petrels. The smallest storm petrels lay eggs that may weigh 29% of the female's weight. That is roughly equivalent to a 20 kg human baby. Whilst women readers will surely shudder at the thought of giving birth to a 20 kg baby, they would have every right to expect their husbands and partners to play a full role in tending the (monstrous) newborn.

So it is with seabird clutches. Once laid, the eggs must receive continuous protection and warmth, and it normally falls to the male to take the first significant turn of incubation duty shortly after laying. This is simply because such a substantial proportion of the female's body reserves have been put into egg formation that she lacks the reserves needed to sustain her over an incubation stint of any length. She must head out to sea to replenish those reserves or, in a few species such as some terns, rely on the male to bring food to her while she continues to incubate.

With the male normally taking the first major stint on the egg, the stage is set for the remainder of incubation. Male and female alternately take turns to protect the eggs, while the partner is at sea feeding, becoming plumper, and building up the body reserves that will allow him or her to take over egg duties several days or weeks hence.

But within this general pattern lies intriguing variation. In some species, particularly those feeding close to shore and probably close to the colony, the time each bird spends incubating before relief by the partner is quite short. Herring Gulls rarely sit longer than five hours. The journeys possible in that time cannot be extensive.

The larger auks, such as Razorbills and guillemots, change over about once a day, but sometimes sit continuously for up to two days. They evidently don't have the time to fly very far from the colony and fit in feeding. This is confirmed by the latest findings. When a team led by Tony Gaston of Environment Canada attached GPS devices to Brünnich's Guillemots incubating on Coats Island, at the entrance to Hudson Bay, they found the average maximum distance the birds flew from the colony was 20 and 27 km in two different years, barely half an hour's flying. Birds from Digges Island, where the colony was ten times the size, flew on average 96 km, about four times further than the Coats Island birds, possibly because the greater concentration of hungry birds on Digges had depleted food close to the colony and generated a halo.[8]

Support for the halo idea also comes from a study of Razorbills, close relatives of Brünnich's Guillemots, lead by Akiko Shoji of the OxNav group. Attaching a GPS logger and a time-depth-temperature recorder to her subjects, she discovered that the further the birds went from the Skomer colony during incubation and chick-rearing, the stronger was the evidence that they encountered good feeding patches.[9] One, but by no means the only explanation, is that the good patches close to Skomer had been exhausted.

Northern Gannets have incubation stints of a similar duration to Razorbills and guillemots. Gannet enthusiast Bryan Nelson recorded that the average length of time for which a male British gannet sat was 36 hours as compared to 30 hours for a female.[10] That allows time for the gannets to fan out from the colony, but little more. These gannet spells are short compared to the shift lengths of Red-footed Boobies, which average some 60–70 hours. This contrast is no surprise since breeding gannets are surrounded by the productive fish-rich summer seas of northern temperate latitudes. In contrast, Red-footed Boobies are tropical. They are light; witness their ability to nest in trees unlike most ground-nesting gannets and boobies. They can fly at comparatively low cost across barren tropical seas, probably ranging up to 150 km from the colony until they encounter small fish driven to the surface by, for example, hunting tuna. Then their agility comes to the fore. Not only do they dive into the sea like other boobies; they are also able to snatch flying fish at the sea surface or even catch them in mid-air.

This has not been a comprehensive survey. Rather it is a brief *hors d'oeuvre* in the build-up to the two groups of seabirds where incubation stints can be seriously long, the penguins and the petrels. Here the sitting bird can remain incubating ashore for a few weeks and its partner has time enough to wander thousands of kilometres. But does it avail itself of that time to travel great distances, or does it simply meander in the vicinity of the colony until it has gained sufficient weight for a lengthy spell of incubation, at which point it sets a homeward course? Today's tracking devices have the potential to reveal the answers.

Among the petrels and albatrosses, there is a weak correlation between size and length of incubation stint. With larger body reserves, the larger species can sit for longer. Average shifts of small storm petrels (2–5 days) and diving petrels (1–2) days are only moderate. Nonetheless one species, Leach's Storm Petrel breeding on Country Island off Nova Scotia, deserves a special mention. Thanks to geolocator information, we can only marvel at a 45 g bird that, in the course of a 6-day off-duty spell, travels a circuit of 2,660 km and reaches an average maximum distance of 1,090 km from the colony. That is sufficient to take the birds into deep oceanic waters south of Newfoundland's Grand Banks.[11]

As expected the shift lengths of the albatrosses are longer, typically 7–20 days, with the all-time endurance record held by a Laysan Albatross which sat, presumably stoically, for 58 days. Satellite tracking has revealed the distances travelled by albatrosses when off duty. These range from mostly no more than 300 km by Macquarie Island's Black-browed Albatrosses, to an unremarkable 1,200 km attained by South Georgia Grey-headed Albatrosses, to a more impressive 2,400 km achieved by Laysan Albatrosses breeding in the north-western Hawaiian Islands. The difference in distances at least partly arises from the differences in productivity of the regions. Albatrosses nesting on South Georgia have productive waters close at hand allowing rapid replenishment,[12] while the Laysan Albatrosses nesting in the far north-west of the Hawaiian chain have to traverse barren local seas before feeding in more nourishing regions in an arc north-west to north-east of the colony.

Arguably the most impressive forays during incubation are undertaken by the mid-sized petrels and shearwaters. Take the case of the Great Shearwaters nesting in literal millions on Nightingale Island and Inaccessible Island next door to Tristan da Cunha in the central South Atlantic. As permanent evidence of their abundance on these islands I recall rocks that are visibly scoured. For millennia they have been scraped by the birds' sharp claws when the 800 g shearwaters scramble for a take-off point above the tussocky vegetation. Once on the wing, eight satellite-tagged males, taking their first break from incubation, provided track-data to Robert Ronconi.[13] Four headed south, up to 1,500 km from the colony, and four headed west, approaching the east coast of South America 4,000 km from the colony. Thus these shearwaters with a body plucked of feathers, barely bigger than a standard jam jar, attained a mean distance from the colony of about 2,500 km.

The longest journeys yet recorded by any albatross or petrel are those undertaken by one of my favourite birds, Murphy's Petrel, a species already met. Velvety grey of plumage and docile of temperament, they are the size of a small gull. They nest on the ground amid low scrubby beachback vegetation on some of the remotest islands of the South Pacific around the latitude of the Tropic of Capricorn (23.5°S), and they are charmingly cooperative. Faced with a curious scientist, an incubating Murphy's Petrel eschews the justifiable aggression of so many seabirds, and simply nibbles the hand that feels under its body to check its legs for the presence of a metal ring, or to briefly remove the egg for measurement.

Studies I completed in the early 1990s on the World Heritage Site of Henderson Island, in the Pitcairn Islands, showed that the average length of an incubation stint was 19 days. This means that complete incubation of some 50 days was often completed in three stints, the male's first followed by the female's, with the egg then hatching during the male's second stint. Wondering where the birds might have gone during their long absences was an inevitable feature of campfire conversation. Could they go as far as California, around 7,000 km to the north, where birds were sometimes seen offshore in June, in the middle of incubation? But it was not until 2011 that I had the opportunity to put geolocator devices on 25 breeding birds to find out. Fortune smiled and we retrieved 18 of the devices two years later.

With the help of Tommy Clay, the output from the devices was downloaded and interpreted. We found a split in the main areas exploited by birds off duty from the egg. Some travelled to low productivity seas about 1,000 km south of Henderson whilst many others went north-east towards the Humboldt Current off Peru and Chile. At the point the birds turned for home, that is to say at the maximum distance from Henderson, they were 3,500 km away on average. That equals roughly the distance across the North Atlantic from Ireland to Newfoundland. The furthest a bird ventured from Henderson on an off-duty excursion was 4,500 km. Impressed by these numbers, we subsequently called on the assistance of our colleague, Steffen Oppel, a dynamic and irrepressible German who works for the Royal Society for the Protection of Birds (RSPB). In 2015 he attached GPS devices to the incubating birds. The total distance covered during the Murphy's Petrels' looping trips towards the Humboldt Current was a remarkable 15,000 km.

Although the petrels' huge journeys to prepare for incubation should put the lid on human grumbles about Saturday morning shopping to fill the larder for the weekend, it is not obvious why they go where they go. The area targeted by the birds from Henderson is merely moderately productive. Perhaps the extra 1,000 km needed to reach the immensely rich seas of the Humboldt current would be a flight too far. Or perhaps Murphy's Petrels are unable to compete among dense milling aggregations of birds at food hotspots, and fare better where food is scattered and can be exploited solitarily and relatively economically thanks to their buoyant flight. At least the birds minimize the energetic costs of the great journey by generally travelling along an anti-clockwise loop, which provides the benefit of following winds as they head eastward on the outward journey and westward when homebound on a more northern track (see also page 120).

Contributing to the vast journeys of Murphy's Petrels is the scarcity of food in the blue emptiness of the tropical South Pacific surrounding their nesting island. However, species breeding near more productive waters also sometimes fly huge distances from the colony during incubation. A nice example comes from the Northern Fulmars breeding on Eynhallow. During incubation, off-duty birds typically remain within 100 km of this Orkney colony or perhaps venture further into the North Sea. But in 2012, one male flew 2,500 km west to the Mid-Atlantic Ridge.[14] He foraged over productive areas of persistent thermal ocean-surface fronts along the Ridge's Charlie-Gibbs Fracture Zone, and travelled over 6,200 km in 14.9 days before returning to relieve his mate. Looking at his track, it is difficult not to suspect that he knew where he was heading as he left the Orkneys! (See Map 5.)

The penguins are the second group that endure long incubation spells. Incubation stints of two to three weeks are the norm in several species. But, for the obvious reason that swimming is much, much slower than flying, off-duty penguins travel but a small fraction of the distance travelled by petrels. For example, ten male Rockhopper Penguins tracked by satellite from colonies in the north-east of the Falkland Islands headed north-east away from land.[15] Three concentrated their foraging over the slope of the Patagonian Shelf about 150 km from home and returned in two weeks. The other seven went further north, up to 350 km from the colony, to which they returned after some three to four weeks. The mean travelling speed on these trips was close to 4 km/h, around the speed of an Olympic swimmer. Another Falklands study confirmed that males made long trips but discovered that females were much more likely to make short trips, and indeed not even stay at sea overnight (which, arguably, shows good sense). Katrin Ludynia of Germany's Max Planck Institute for Ornithology thought the difference might be due to the fact that males needed to build body reserves during incubation so as to be fat and fit to guard the newly-hatched chick. In contrast females had less need to build up reserves since their role of collecting food for and feeding the small chick also allowed them to feed themselves at this stage.[16] Here then is an example of the different responsibilities of the two sexes being reflected in different seagoing journeys.

I cannot resist pursuing that theme and ending this section by recounting what modern studies have revealed of the travels of the off-duty Emperor Penguin, the female. Remember that the female lays her single egg onto her down-turned tail in June at the start of winter. Then, anxiously, she transfers it onto the feet of her mate who tucks the valuable cargo under a cosy fold, and prepares for the winter ordeal. He huddles up to his fellow males and endures the worst of the Antarctic winter. Apsley Cherry-Garrard entitled his 1922 book from the heroic age of Antarctic exploration *The Worst Journey in the World*. Although his journey to collect an Emperor Penguin egg was a hell of frigid endurance, it was no worse than the winter conditions, down to -40°C, that the male Emperors routinely withstand while their mates are away at sea. The females only return when the chick is due to hatch some 64 days after laying. But where do they go during this two-month break?

Fitted with satellite transmitters or time-depth recorders by Roger Kirkwood and Graham Robertson, female Emperor Penguins left the colony on the coast of the Australian sector of Antarctica.[17] They first walked and tobogganed for 80 km across the fast ice to reach open water. The better part of two months was then spent in polynyas, more or less permanent openings in the sea ice, within 100 km of the colony. Here the birds hunted food in water 200–500 m deep, over the outer continental shelf or shelf slope where the seafloor begins its descent to the deep-ocean abyss. Occasionally, when in transit and when on the ice

Incubation duties completed, male Emperor Penguins head across the ice to open water, leaving their mates to tend the newly-hatched chick.

between foraging bouts, the females huddled together to minimize heat loss, exactly like their mates at the colony. Life was evidently as grim for them as it was for their mates.

During interludes in incubation some seabirds cover truly amazing distances. Where the distances are large, say greater than 500 km, it is very evident that the birds do not fan out equally in all directions from the colony. Rather, they head in specific directions, pass more or less rapidly through vast tracts of ocean that presumably offer little, and then concentrate their foraging efforts in certain defined oceanic regions. The features of those preferred regions are explored in Chapter 8.

* * *

Once the eggs have hatched and the chicks have stretched their legs and stubby wings after that long period of enclosure, the routines of the parents change. Initially the change may not be especially abrupt. Among

**PLATE 1**. Where the Atlantic Puffins of Great Britain spent the winter after quitting their colonies was unknown until their remarkable peregrinations were revealed by modern tracking. Geolocators, as attached to the top bird, have been very informative. (Bottom picture ©jamiegundry.com. Top © Dave Boyle.)

**PLATE 2**. The Rockhopper Penguins of the Falklands can travel up to 350 km from the colony when feeding at sea while the mate tends the eggs (© Michael Brooke).

**PLATE 3**. The long-distance movements of Common Terns breeding in Germany can be tracked by geolocators on the bird's legs (upper), while comings and goings at the colony are monitored by an antenna surrounding the nest that detects the bird's implanted PIT tag (lower). (Upper © Sabrina Weitekamp, Lower © Peter Becker.)

**PLATE 4**. Although the frigatebird's tactic of harassing Masked Boobies and other seabirds is conspicuous to onshore birdwatchers, modern tracking has shown how frigatebirds remain in continuous flight far offshore for periods of a month or more. (Both © Michael Brooke.)

**PLATE 5**. Upper: A juvenile Wandering Albatross prepares to take its first flight that will be tracked via the satellite transmitter attached to the youngster's back (© Richard Phillips). Lower: Commonly smaller than satellite transmitters, geolocators, here attached to the leg of a South Polar Skua, have revolutionised knowledge of bird migration (© Richard Phillips).

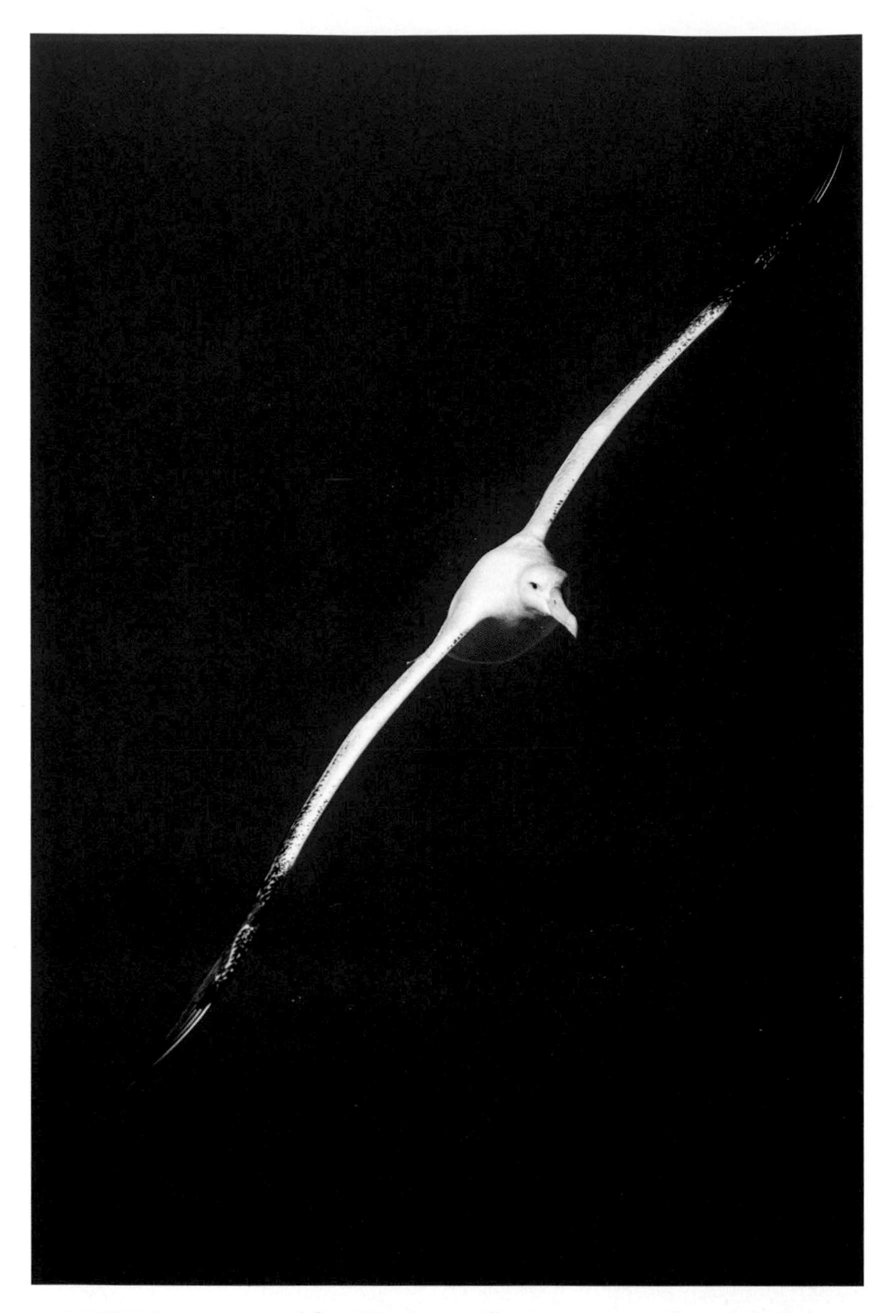

**PLATE 6**. It is not unusual for a Wandering Albatross to pass its thirtieth birthday, by which time it will have glided for millions of kilometres (© Oliver Krüger).

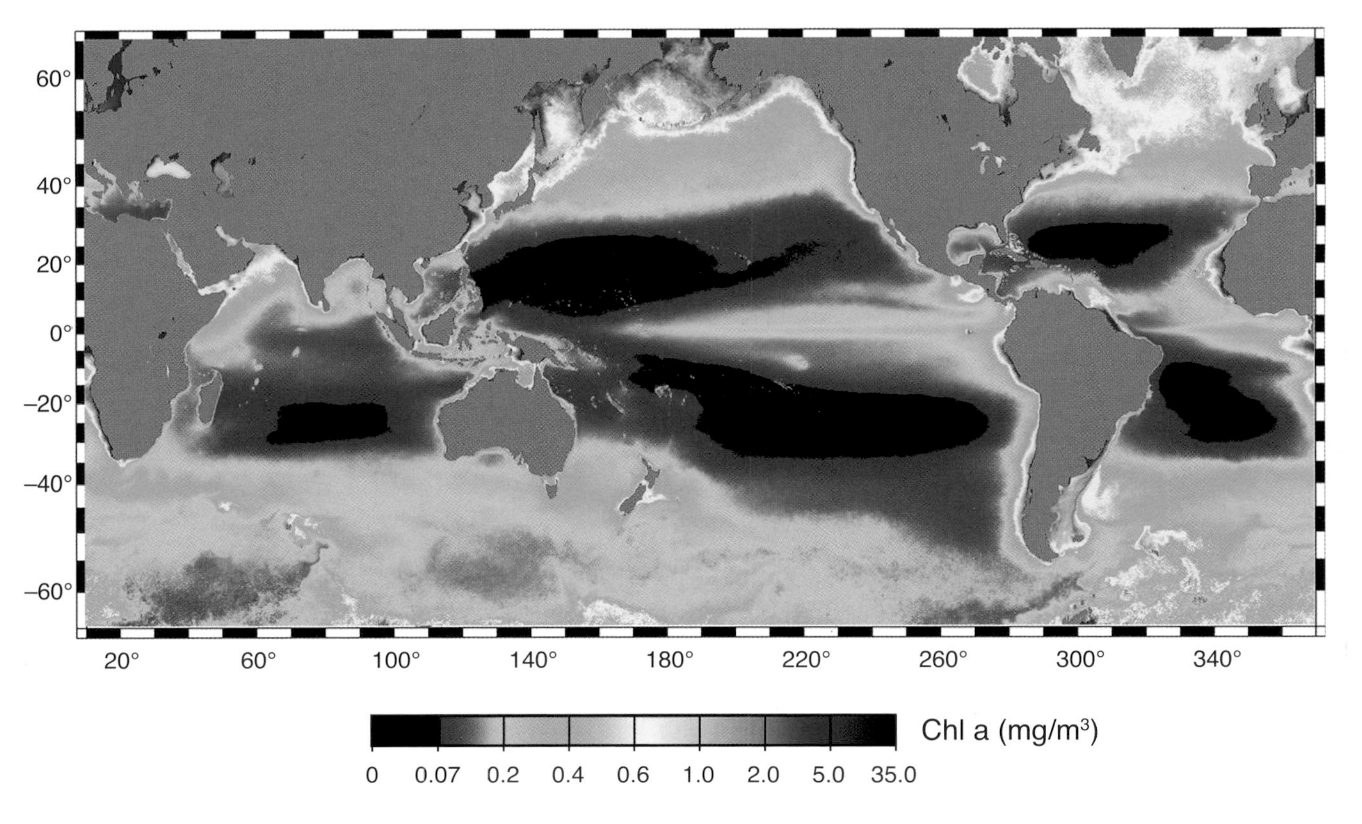


**PLATE 7**. The image shows marine productivity, as revealed by chlorophyll concentrations, in the world's oceans. Generally, productivity, measured by $mg/m^3$ is highest (redder tones) at higher latitudes and along continental margins. It is lower (bluer tones) nearer the Equator, and conspicuously low (black) in mid-ocean along the Tropics of Cancer and Capricorn (Source: NOAA).

**PLATE 8**. The vast majority of the world's Laysan Albatrosses breed in the Hawaiian Islands, from where their travels have been well-studied (©Jacob González-Solís).

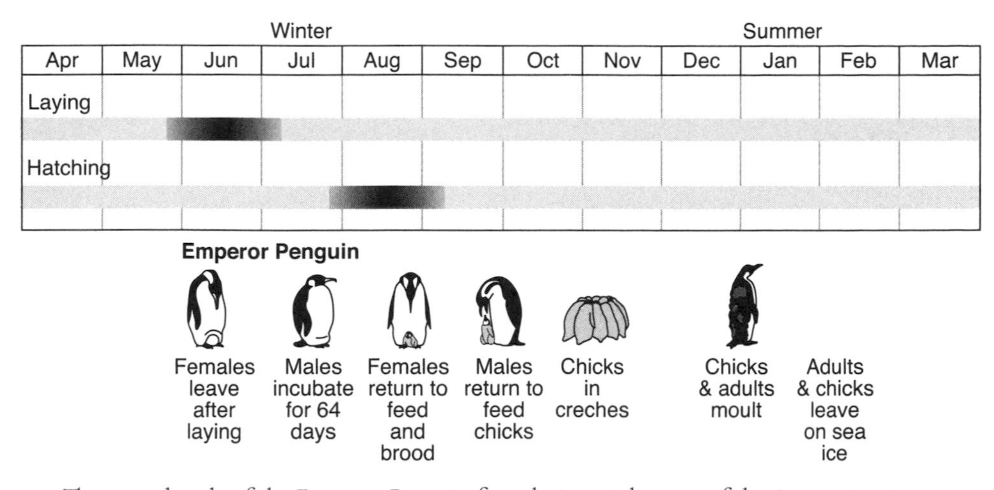


The annual cycle of the Emperor Penguin from laying at the start of the Antarctic winter in June to the chicks' departure in February.

many species, the youngest chicks require constant brooding and so one parent remains at the nest to provide the chick with warmth and attention whilst the other is at sea collecting food. After that so-called guard stage, which lasts from a few days to some three weeks according to species, the chicks are left alone while the parents are away at sea searching for food to deliver to the nest.

Given that seabirds with chicks return to the colony every few hours or every few days, it is relatively easy to place a tracking device on them to establish where, for example, the Puffin has caught that beakful of sand eels or the Lesser Black-backed Gull has obtained that mushy ball of earthworms. Studies on numerous species have demonstrated that birds with chicks, to which they must return more or less frequently, feed closer to the colony than they do during incubation when, as we have seen, the absences and journeys can be lengthy. To try to make sense of the torrent of information that has accrued, I shall shirk a comprehensive account. It would be over-lengthy. Instead I will explore some influences on parent seabirds foraging for their chicks.

A very obvious possible influence is the age of the chick. Many years ago I studied Northern Wheatears, small territorial songbirds that enjoy open country. I wondered if those adults with small chicks, needing relatively infrequent visits, fed further from the nest than adults with large chicks. I found that they did.[18] This might have twin merits: it would reduce the risk of attracting predators to nests with small chicks, and leave food close to the nest to be used when the older offspring were at their most demanding. The second advantage only comes to pass if the wheatears successfully defend their territory against other wheatears – as they generally do. It seems far less likely to work for a seabird where, as far as is known, no individual has its own defended and exclusive feeding area.*

This leaves several possibilities for a seabird. The pedestrian possibility, the null hypothesis in science-speak, is that the parent's foraging range alters little with chick age. Alternatively the foraging range could expand for one of several reasons. During the brood phase, the parent might remain close at hand. Later, when the chick can be left for longer periods and gulp down larger meals at one time, the parent might forage further afield. This trend would be reinforced if food were depleted near the colony. Finally, a parent might collect food for its chick(s) quite near the colony, minimizing the transport costs, but go further afield to feed itself where competition from its fellows is probably lower. This strategy, effectively leaving chick food close to the colony, bears comparison with the wheatear idea.

Satellite tracking has shown albatrosses to remain close to the colony during the brood-guard phase. For instance, the Black-browed Albatrosses of South Georgia mostly feed north of the colony during incubation, the females somewhat further away (1,700 km) than the males (1,000 km). For both sexes, these ranges contract to a mere 300 km during the brood-guard stage, before expanding again once the chick can be left unattended between feeds.[19] Other albatrosses, such as the Wandering Albatross, show a similar pattern; birds are particularly concentrated near the colony during the brood-guard stage. To protect the birds from tragic interactions with fishing vessels, there is a strong case for closing the waters around the colony to fishing during this key period, as has been done around South Georgia.

* See page 106 for evidence that adjacent colonies have distinct feeding areas, albeit ones that are probably not defended in the way that nest sites or territories are.

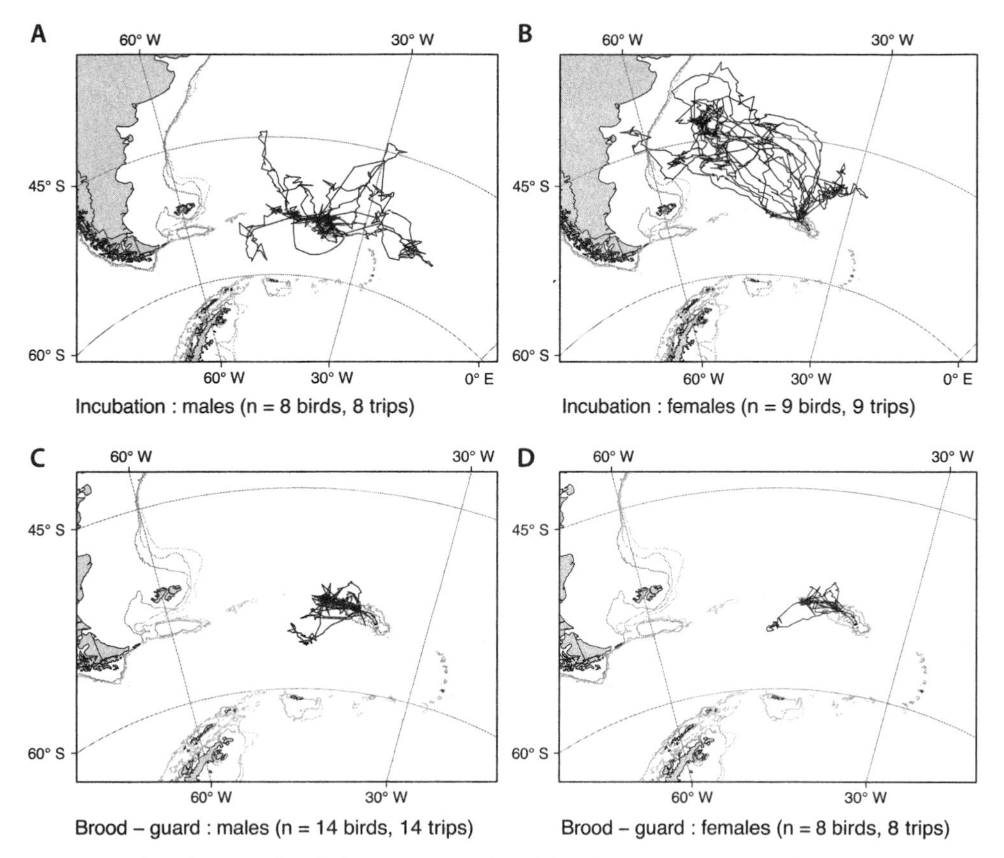


The distribution of male (map A: 8 trips) and female (map B: 9 trips) Black-browed Albatrosses tracked from South Georgia during incubation is clearly different. However, during the brood-guard stage at the start of chick-rearing, the trips are much shorter and barely differ between males (map C: 14 trips) and females (map D: 8 trips). Reproduced with permission of The Royal Society, from the work cited in Note 19, Chapter 5.

Only in the 1990s was it realised that parent albatrosses and petrels feeding their chicks may alternately embark on short feeding trips lasting 1–3 days and longer trips lasting over about five days. This habit has proven widespread, but not universal. It can entail a simple alternation of short and long trips, or a more complex pattern, the parent making 2–5 short trips before embarking on a long one. In general, the short trips result in a greater rate of food delivery to the chick, but at a cost to the parent. The parents lose weight which is regained on the long trips.

That much can be learnt by observing the coming and goings of birds at the colony. It does not illuminate whether the two trip types involve different destinations.

Two nice examples, confirming that short and long trips do indeed involve different destinations, come from the Southern Ocean. The nesting stronghold of the Sooty Shearwater is New Zealand where unlucky parents may raise a chick that becomes a harvested muttonbird served in a New Zealand restaurant.* When adult shearwaters with chicks were tagged with geolocators, they mixed short 1–2 day trips over New Zealand waters, typically within 500 km of the colony, with 11–14 day trips to the far colder seas of the Antarctic Polar Front 2,000 km south from the colony.[20]

In terms of distance ventured the contrast between short and long trips is even more extreme among the White-chinned Petrels nesting on the Iles Crozet, the archipelago in the Southern Ocean. On short trips lasting just over one day, the petrels' average maximum distance from the colony was 62 km, barely over the horizon. On long trips averaging 8.5 days, they matched the Sooty Shearwaters and headed south, a fast commute averaging 31 km/h to Antarctic waters almost 1,900 km distant.[21] The homeward journey from these krill-rich seas was equally rapid at 34 km/h.†

In the majority of seabirds, males and females are of similar size and have equivalent roles in chick-rearing. That said, there are exceptions. When the female Emperor Penguin returns to the colony to relieve her mate after his display of winter fortitude caring for the egg, she guards the newly-hatched chick for the next 3–4 weeks. Conversely, in Rockhopper Penguins, the males are primarily responsible for guarding small chicks.

Later in chick rearing, less expected gender differences have been discovered. Boobies follow the relatively uncommon pattern of size dimorphism also seen among hawks, skuas and owls. The female is the larger, which may contribute to differences in sexes' behaviour. In the

* Rich, gamey, and delicious in my single experience.

† That is just slower than Usain Bolt at full tilt, remembering that a 10 second 100 metre dash equates to 36 km/h.

Masked Booby, there is also a particularly striking difference in call; a bold honking by the female and a plaintive whistle from the male. It is she who delivers more food to the chicks. In the Brown Booby, a species whose clean-cut chocolate-brown and white plumage is a triumph of graphic design, the larger female makes longer journeys than her mate when they are searching for sardines in the Gulf of California, albeit journeys that would barely make a petrel sweat. The female's journeys last for three hours, rather than the male's two, and take her 40 km from the colony, rather than his 17 km.[22]

A sexual contrast of a very different nature has emerged from studies of Southern Hemisphere shags. It has been best documented among the Imperial Shags of Argentina where males are about 18 percent heavier than females. With poorly-waterproofed plumage, this is a species that spends the night ashore. One might guess that birds would first enter the water around dawn and finally come ashore around dusk. This is exactly what birds, studied in southern Argentina via geolocators and immersion loggers, routinely do outside the breeding season. But, during incubation and early chick-rearing, the females enter the water at dawn and come ashore at midday. That is the cue for the males to go into the water, from which they finally emerge at dusk. For part of the year, it is a clear instance of 'ladies first'.[23] Assuming this difference is not the result of masculine politeness, the study's authors struggle to explain it. Possibly the larger male can better defend the nest, and more readily adjust his afternoon feeding schedule if his mate is late returning after an unsatisfactory morning.

There is something preposterously self-important about a male frigatebird inflating his red throat pouch. But it works. It not only attracts a mate but, unusually among seabirds, she is a mate who tolerates his steady withdrawal from chick-feeding which may last (for her) a remarkable 15 months.* Indeed differences in the chick-feeding behaviour may begin as early as the brooding period. Studying Christmas Island

* This includes feeding after the chick has fledged; see page 35. Normally, as mentioned in Chapter 4, seabirds retain the same mate year after year. The fact that the female frigatebird cares for the chick for far longer than does the male doubtless contributes to their failure to follow standard seabird practice. Among frigatebirds successive breeding attempts are with different partners.

Frigatebirds on the eponymous Indian Ocean island, famous for its marching crabs and notorious for its Australian Immigration Detention Centre, a French team deployed satellite tags and GPS trackers. In 2009, there was no difference between the foraging journeys of males and females in the brood period but, in 2010 when feeding conditions were poorer, the male frigatebirds' trips lasted twice as long as those of females, and took the birds further from the colony (413 v. 194 km).[24] It was almost as if the males were already easing out of any long-term commitment.

As a generalisation, a parent seabird collecting food for its young should strive to minimize transport costs and collect that food as near to home as possible. Conversely, when the going gets tough, the tough seabird must face going further. That line of thinking could apply at several time scales. Over a period of days, local seas might be influenced by weather and force the bird further afield. Across years, birds may feed closer in one year and further in another, possibly poorer, year. If the deterioration is long-term, perhaps because of climate change, a colony could become untenable. Finally, because of the Ashmole halo, there is the possibility that birds from larger colonies habitually feed further afield than their counterparts at smaller colonies.

Chris Feare is a lucky man. Now in his seventies, he has long since run out of fingers on which to count the number of times he has visited the luxury ecolodge* on Bird Island, Seychelles. The island hosts half a million pairs of Sooty Terns, also known as Wideawake Terns in celebration of their unrelenting 'wideawake' shriek. When incubating birds were GPS-tracked by Feare in 2014, he witnessed short-term changes in foraging journeys. The terns' off-duty shifts initially lasted 1–2 days, during which the birds covered 100–200 km. Then there was an abrupt change to 6-day off-duty spells when the birds scoured 2,000 km of ocean. This incident coincided with a brief spell when sea surface temperature dropped in the western Indian Ocean. Chris Feare did not know whether this temperature fall affected the birds' food supplies directly or, alternatively, influenced the movements of tuna upon which Sooty Terns depend to drive their small prey to the sea surface, and into

* http://www.birdislandseychelles.com/ (accessed 14 June 2017).

the beaks of the terns gathered overhead. Whichever was correct, the outcome was that many terns abandoned their eggs.

Not only do short-term fluctuations influence how far seabirds need to venture, so too do changes in feeding conditions between years. Tracked by GPS from Coquet Island, off England's beautiful Northumberland coast, Black-legged Kittiwakes embarked on markedly different trips when feeding chicks in 2012 as compared to 2011.[25] The 2012 trips were far quicker (averaging 2.9 h versus 5.1 h) and shorter (averaging 20 km total journey length versus 64 km), apparently reflecting differences in the availability of sand eels, their prey-of-choice.

If parents caring for chicks need to forage further from the colony in lean periods, we might guess they would also be constrained to travel further afield in regions where the ocean is unproductive. Depending on offshore conditions, there can be a ten-fold difference in the distance British Common Guillemots at different colonies fly to hunt for food for their chicks. But a particularly nice example comes from southern Argentinian Patagonia where Magellanic Penguin colonies are dotted along 1,000 km of dry coast. Roughly, the further south the colony is situated, the more productive are the waters offshore. Dee Boersma of the University of Washington found that male penguins feeding chicks in the south travelled a maximum distance of some 100 km from the colony. Their northern friends travelled twice that distance.[26] Furthermore, at colonies where Magellanic Penguins in this region have to work harder on a foraging trip, measured as trip duration, total distance covered and maximum distance from the breeding colony, the population is least likely to be growing.[27]

Quite often it will be tricky to decide whether birds from different colonies forage at lesser or greater distances because the surrounding seas are more or less productive or because a multitude of hungry mouths has depleted fish stocks near the colony (the aforementioned Ashmole halo effect). The Magellanic Penguins are an example of the former. A study by Steffen Oppel is a probable example of the latter. Oppel studied breeding Masked Boobies on Ascension Island in the tropical Atlantic Ocean and on St Helena, site of Napoleon's last exile and death, 1,200 km to the south-east.[28] The waters surrounding both islands are equally unproductive, the blue water 'deserts' of the tropics. But, including other species targeting the same prey as boobies, the total number of seabirds on Ascension, around 25,000, was 50 times greater than on St Helena, around 500. Whether during incubation or chick-rearing, the GPS-tagged Ascension boobies undertook journeys that were around twice as long in distance and duration as those of the St Helena birds. However, the 'idler' St Helenan boobies had the higher breeding success, likely because there was more food close by the smaller colony.

If the birds are influencing food supply in the vicinity of their colony, it would make sense if colonies were spaced out across the available islands, headlands, and cliffs. And more than 30 years ago, even before the invention of GPS, Bob Furness and Tim Birkhead, respectively hailing from the Universities of Glasgow and Sheffield, reported as much.[29] They found that, for four British seabird species, larger colonies were surrounded by larger zones more or less bereft of other colonies of the species than were smaller colonies.

While this implies a tendency for the birds from each colony to forage in a colony-specific zone, no-one anticipated the extraordinary results that emerged when Ewan Wakefield of Leeds University co-ordinated tracking of 184 chick-rearing Northern Gannets from 12 colonies dotted around Great Britain, Ireland and northern France.[30] The birds' foraging tracks fanned out from the colonies. So far, so expected. But the lack of overlap of the tracks of the birds from neighbouring colonies was startling. It was almost as if the birds had encountered a mid-ocean 'Trespassers will be prosecuted' sign, and obediently turned around. (See Map 6.) Most striking was the pattern off the west of Ireland. There are situated two colonies, Little Skellig (29,700 pairs) and, about 30 km to the south, Bull Rock (3,700 pairs). Little Skellig birds head overwhelmingly north-west, away from neighbouring Bull Rock (and the Irish mainland), while the Bull Rock birds head south.

Just how this segregation of birds between colonies arises remains mysterious. Ewan Wakefield speculates "We wondered whether birds, at least partly, learn the best feeding areas by following other gannets leaving their own colony. If this happened over several gannet generations, a tradition of avoiding the areas exploited by the neighbours, the gannets from other colonies, could build up." The idea gains credence from Wakefield's observation that immature gannets follow adults departing the colony. Compared to their elders, the immatures have less knowledge of prime feeding areas, but perhaps have more 'spare' time to learn the whereabouts of those areas. This line of thinking matches the finding (Chapter 7) that adults are the more consistent in their use of a particular feeding area.

A persistent message from this chapter has been that birds tend to forage further from the colony during incubation than chick-rearing. If colonies tend to be spaced apart - as is the case - that would automatically lead to birds from different colonies experiencing less overlap of foraging areas during chick-rearing. Over three breeding seasons, the OxNav group studied Manx Shearwaters on Skomer Island, at the southern margin of the Irish Sea, and on Copeland Island, off Belfast, just over 300 km to the north.[31] GPS tracking revealed much overlap during the incubation period in a rich feeding area, the Irish Sea Front, southwest of the Isle of Man, and much closer to Copeland than to Skomer. Overlap was reduced during chick-rearing when birds were feeding closer to home.

In other species, there is more pronounced mingling between the birds from neighbouring colonies. An example is described in a paper with the give-away title 'Social foraging European shags: GPS tracking reveals birds from neighbouring colonies have shared foraging grounds'.[32] Since the three colonies studied, in the Scilly Isles, south-west England, were separated by short distances, only 4-10 km, there was probably opportunity for the birds from different colonies to share information about where to feed. Nevertheless the shags, not the best fliers, did most of their foraging within 3 km of home, rather than in shared seas.

If birds of the same species from different colonies tend to avoid sharing feeding areas, it is intriguing to ponder whether or not birds of different species sharing the same breeding station share feeding niches. Classical ecological theory would suggest that there is likely to be some segregation of feeding niches. For example, one species may feed deeper underwater or further from the colony than another, or may simply catch different food. That said, the interspecific differences generated by competition may be slight compared to the differences in feeding routine that I have described for different stages of a species' breeding cycle.

Where the species under comparison are very different, the question of feeding segregation is meaningless. South Polar Skuas patrolling an Antarctic Petrel colony to grab petrel eggs and chicks are of course foraging in a totally different way to the krill-eating petrels. The question gains interest when two species breeding alongside each other are more similar. Consider Macaroni and Rockhopper Penguins. The former is almost twice the mass of the latter but both are crustacean specialists of the open ocean, the Southern Ocean, that happen to be adorned with colourful head crests and frequently share breeding stations. One such is Marion Island, four days steaming south-east of Cape Town, where the penguins were studied, via GPS tracking and time-depth recorders, by Otto Whitehead of Cape Town University's FitzPatrick Institute of African Ornithology.[33] Both during chick feeding and in the subsequent build-up to moult the larger Macaroni Penguins dive deeper and for longer than the Rockhoppers. They also tend to venture further from Marion.*

Favoured for studies of this ilk are the larger auks. This might be anticipated. In the North Atlantic Razorbills, Common Guillemots, and Brünnich's Guillemots nest in clamorous colonies, often together on guano-smeared cliffs. Each species appears to bring the single chick rather similar prey. How, if at all, do they divide up the available resources? Consistently, studies report that Razorbills dive to lesser depths than do the Guillemots (e.g. to 12 m as compared to 30–35 m for the Guillemots of both species breeding off south-west Greenland[34]). This is almost certainly related to the fact that Razorbills possess larger wings in relation to body mass; good for flying but almost a hindrance underwater. This has two further consequences. Razorbills may fly further from the colony, at least when feeding chicks at a Baltic Sea colony, and they appear to undertake V-shaped dives, catching fish on the ascent, while Guillemots use their greater diving skills to search for and catch prey at the bottom of their U-shaped dives. These contrasting dive profiles are considered in more detail in Chapter 9.

* During the month-long pre-moult excursion, maximum foraging ranges of Macaroni Penguins averaged 903 km, greater than the Rockhopper average of 696 km.

When seabirds are breeding and tied to a colony by their parental duties, how far afield they forage is clearly influenced by a variety of factors. Some of those factors are intrinsic to the birds, in particular the stage of the breeding cycle. Others are external to the birds. While a prudent strategy might be to feed as close to home as possible, this may not be achievable if feeding hotspots are further afield, or if fellow colony members have depleted the local food supply, or if other species more or less monopolize local food. These factors place a premium on travelling further afield as economically as possible. The tactics seabirds use to reduce their travel costs during breeding, and indeed at other times of year, are slowly becoming understood.

# Wind and Waves

## Friend and Foe

Cut off from a raft, a boat, or a larger vessel, a naked human in mid-ocean is unimaginably helpless. In the tropics, imagine the lingering terror of wondering when the sharks will arrive. In the icy seas of the North Atlantic, any terror will be short-lived as the body quickly becomes chilled and the brain shuts down within minutes. Yet tropical seas, temperate seas, and icy seas provide the very conditions with which seabirds routinely cope, be it a penguin passing half a year out of sight of land, or a Sooty Tern spending months on the wing, never alighting on the blue tropical sea, never ceasing to flap its wings. Indeed, not only

Grey-headed Albatrosses have been recorded sustaining the astonishing speed of 130 km/h, when driven homeward by Southern Ocean storms (© Oliver Krüger).

do seabirds cope, they have tricks that enable them to exploit the conditions they encounter at sea to save energy and prosper. Storm petrels pitter-pattering on the sea surface in the depths of 10 m troughs as mighty Southern Ocean swells tower above are clearly coping, as is the female Grey-headed Albatross rocketing downwind at over 130 km/h, blasted homeward to South Georgia by furious tailwinds of 70–80 km/h.[1]

Biologists routinely ask how the behaviours adopted by an animal could help it save energy. The presumption is that energy saved means the creature might, for example, be better able to survive foul weather, or to deliver more food to its chick and enhance the survival chances of the youngster. Thus this chapter will focus on evidence that, at various spatial scales from metres to thousands of kilometres, seabirds adjust their journeys to minimize travel costs. The separate but related question of whether they select the most profitable feeding areas to visit will be the focus of Chapter 8. Inevitably things sometimes go wrong. There are occasions when a storm strikes and the seabird becomes embroiled in the maelstrom of rain, salt spray and driving wind, and finds itself who knows where. Tales of such seabird 'wrecks' will end the present chapter.

For the very obvious reason that they cannot fly, penguins live their lives on a smaller spatial scale than, say, albatrosses. Nonetheless, as we have seen in earlier chapters, the scale can be hundreds or even thousands of kilometres. A long way to swim, and swimming requires more energy per kilometre covered than does flying. One feature of penguin swimming, very obvious to anyone who has seen them onscreen or in the flesh, is the habit of frequently leaping clear of the water in the manner of a porpoise. For porpoises and dolphins, the behaviour, characteristic of higher speeds, has been suggested to save energy. This is because, at higher speeds, the drag associated with swimming is higher when the bird is within three body diameters of the surface than when it is swimming at greater depth. In addition, the extra time above the surface due to leaping clear could facilitate a necessary gulp of air - and note that fish, which do not breathe air, do not porpoise.

As so often in science, a story which seemed secure wobbled when new study techniques could be brought to bear. This has happened to the energy-saving explanation of penguin-porpoising. Japanese researcher Ken Yoda, now of Nagoya University, is bold enough (or rash enough?) to assert on his personal website that there are no animals on earth that cannot be tracked.[2] Using accelerometers, his team has been able to document the swimming behaviour of Adélie Penguins at sea. Even in the absence of ice, Adélie Penguins collecting food for chicks spend only about 1 percent of their time at sea porpoising, and only some 4 percent of the total distance swum is covered using this technique. These observations seem to undermine the energy-serving idea, prompting Yoda to wonder whether porpoising, mostly employed near to shore, is actually primarily a tactic for avoiding predators that are most likely to be encountered near to shore.

Penguins could also save energy by travelling in the same direction as ocean currents. The Iles Crozet in the Southern Ocean are a penguin stronghold, and the largest King Penguin colony in the archipelago is home to around half a million pairs. What a tumult of throaty calls and flapping flippers. When breeding, the birds mostly forage to the south near the Antarctic Polar Front and the outward journey is aided by a following current.[3] It is uncertain whether the penguins head south because that takes them to the best feeding areas, or to take advantage of the current. The fact that they can return home into the current without undue difficulty might support the former explanation.

If a penguin porpoising epitomizes the frantic bustle of life, its antithesis is a languid albatross. Wedged towards the stern of a 5,000-tonne research ship, I first saw a Wandering Albatross on a journey from Cape Town to Marion Island, 2,000 km to the south-east. From the grey distance the bird emerged. For perhaps 30 minutes, it planed above the ship's churning wake, hoping for a garbage snack. Not once were its immense drooping wings flapped. But they were continuously adjusted, just as a leading racing driver is forever twitching the steering wheel in an effort to retain the fastest, racing line. The adjustments were most obvious when the bird swooped to near the surface, perhaps even brushing the water with a wing tip.

Sailors and biologists have long been fascinated by the ability of albatrosses to glide for hours with barely any movement of their wings. A key factor is anatomical; a shoulder lock in albatrosses and giant petrels has the effect of reducing or even eliminating the need for any muscle power to hold the wing outstretched and horizontal. This occurs because, when the wing is fully extended, the lock resists any attempt to raise the wing above the horizontal. As soon as the humerus is slightly retracted from the fully forward position, the lock no longer operates, and so the wing can be raised above the horizontal. A related anatomical adaptation is the bulk of albatross flight muscles. They are smaller, relative to body size, than those of smaller flapping species, and the supracoracoideus muscle used to raise the wing during an upstroke is especially small. With less muscle mass to keep airborne, the gliding albatross saves energy.

Albatrosses glide with wings locked for long periods using one of two techniques. The first and more straightforward involves slope-soaring. Wind is deflected upwards from the windward side of waves. If an albatross turns into this wind, it gains height, albeit with some loss of ground speed, just as a paraglider can gain height over land as wind is deflected upwards off a hillside. Having gained height, the albatross can

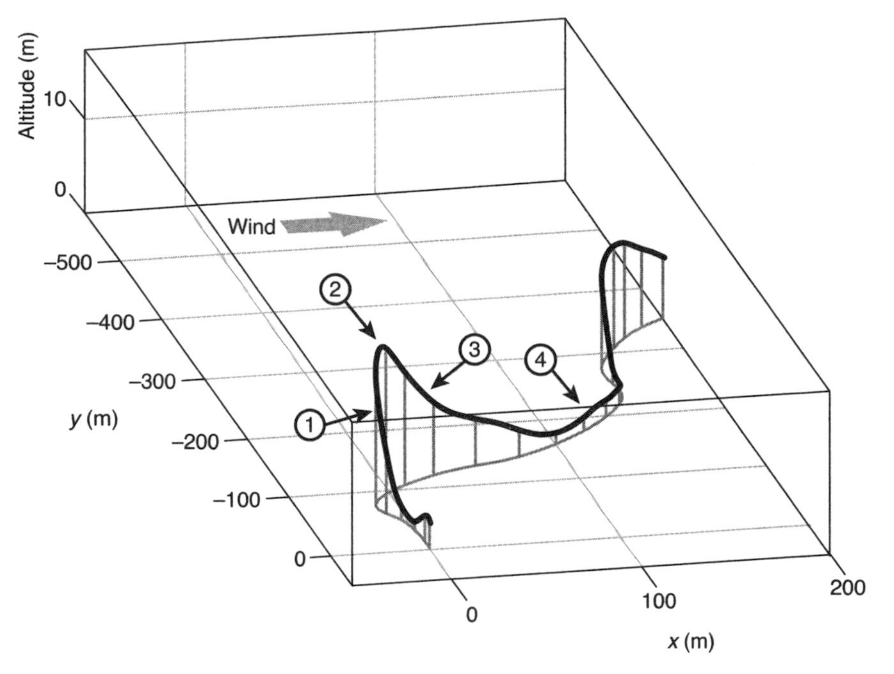


The dynamic soaring cycle of an albatross consists of (1) a windward climb into wind, (2) a curve from windward to leeward at the highest altitude, (3) a leeward downwind descent and (4) a curve from leeward to windward at low altitude close to the sea surface. During a single cycle the bird covers a few hundred metres.

now gather speed as it glides downwards, perhaps roughly at right angles to the wind direction and therefore parallel to the waves, before turning into the wind to gain more height and to repeat the manoeuvre. The net track is, broadly speaking, across the wind.

The second technique is dynamic soaring, manifest in the albatross ascending into wind, turning at the peak of the climb, descending and accelerating downwind before beginning another climb. Such repetitive dynamic soaring exploits the fact that drag slows the passage of air close to the sea. Consequently the speed of the wind is least immediately above the water, increases with height for about 5–10 m, and then nearly levels off. Thanks to this wind gradient, lift is generated when the bird turns into wind to make the climb since air flows over the upper surface of the wing faster than over the lower surface, and consequently pressure is lower above the wing than below. The effect, enhanced by the cambered shape of the albatross wing, is analogous to the generation of lift over an aircraft wing. Then, as the combination of drag, the downward pull of the bird's weight, and the diminishing wind-speed gradient prevent further useful ascent, the bird reaches the phase of the cycle when it has minimum energy. It turns to leeward and glides downward, gaining energy before turning into the wind and beginning the cycle afresh. The overall track is characteristically somewhere between downwind and across the wind. It needs wind speeds of at least 30 km/h, which of course explains why albatrosses are mostly found in the world's windier oceans where they are spared the energetic costs of flapping.

The techniques described rely on the fact that albatrosses are very efficient gliders that can travel a considerable distance horizontally for every metre dropped vertically. Their glide ratios are around 22–23. This means that, on the descent from a height of, say, 8 m, an albatross can travel 22 times that distance, about 180 m, before rising upwards once more; worse than a man-made glider (glide ratio 40–45), but better than an eagle or vulture (about 15). However, the track is necessarily zigzag, and so albatrosses and indeed other petrels are travelling further than the simple straight-line distance between start and end points. In the larger albatrosses, the zigzag track is about 50% further than the straight-line distance.

Languid is the word that comes to mind when watching a gliding albatross. And the efficiency of albatross flight is confirmed by modern heart rate measurements of gliding Wandering Albatrosses. Combining information from externally-attached heart beat monitors, leg-mounted immersion recorders that reported whether bird was afloat or airborne and a satellite pack for relaying the information back to Toulouse in France, French scientist Henri Weimerskirch discovered that the heart rate of flying Wandering Albatrosses, around 80 beats/min, was barely faster than the rate of birds resting on the water, around 60 beats/min.[4] But take-off, involving flapping, was seriously hard work. The heart rate then topped 200 beats/min.

If the Wandering Albatrosses seem well-monitored, spare a thought for Red-footed Boobies, also studied by Weimerskirch. The boobies nest on Europa in the Mozambique Channel from which, over the course of the tropical day, they make 12-hour excursions that take them up to 150 km from the colony. The birds were variously asked to carry GPS recorders, accelerometers, time depth recorders, activity recorders, and altimeters, fortunately not all at once! When travelling between foraging spots, the birds flapped for about one-third of the time, in 1.7 second bursts, and glided for the remainder in 3.8 second bouts.[5] This is likely to be a highly efficient means of progression, judging by data from the booby's near-relative, the Cape Gannet. Monitored from South African colonies via electrocardiograms, the gannets' average heart beat during flapping was 250 beats/min. This dropped almost instantaneously by 13 percent, to 217 beats/min, when the birds entered a gliding phase. And 217 beats/min is barely faster than the heart rate of birds resting on the water, around 210.[6] Like albatrosses, these birds can travel the seas very economically, even when flapping.

In general sustained flapping is more problematical for larger birds than for smaller. Vultures and eagles are not enthusiastic flappers. Instead they routinely soar in updrafts or thermals to gain height and then glide downward while progressing across country horizontally. Remarkably – and this has only been fully appreciated over the past 15 years – frigatebirds adopt similar tactics over the open ocean.[7] Truly bizarre birds with sharply angled wings, frigatebirds cannot safely land on the sea. Their plumage lacks waterproofing and their long wings obstruct take-off. This combination of features is resolved because, at sea, frigatebirds exploit thermals under cumulus clouds in the tradewind belt north and south of the Equator.* Having used these thermals to ascend up to 2,500 m – and very occasionally as high as 4,000 m where the temperature falls below freezing – the frigatebirds can then glide downward to around 90 m before ascending once more. This cycle continues day and night for as long as the frigatebird is at sea. Only sporadically, perhaps half a dozen times in the 24 hours and most probably by day, does a frigatebird come lower, to the sea surface, where there is the prospect of food.

The capacity of frigatebirds to spend days at sea raises one very obvious question. Do they "sleep through all the night with open eye" as

* Along the Equator lie the Doldrums, the zone of light winds so frustrating for sailors and so inimical to frigatebirds.

Chaucer suggested for small birds in the Prologue to the Canterbury Tales? Over 600 years after Chaucer wrote, an astounding answer has emerged for adult Great Frigatebirds – and presumably the answer would be similar for juveniles.

Niels Rattenborg from Germany's Max Planck Institute for Ornithology was able to make electroencephalogram (EEG) recordings of Great Frigatebirds breeding in the Galápagos and flying over the ocean for up to 10 days. The recordings indicated periods of sleep, just as human sleep can be detected by EEG monitoring. At sea, the birds slept in short bursts totalling about 42 minutes in each 24-hour period, mostly at night. This represented major sleep loss compared to the daily total of just over 12 hours achieved when ashore.

More remarkably still, the team was able to obtain independent data from both the left and right hemispheres of the birds' brains. While about half the onshore sleep involved just one hemisphere, this proportion increased to three-quarters at sea. The frigatebirds could then keep one eye open, literally and metaphorically, for hazards. But the fact that one-quarter of at-sea sleeping involved both hemispheres simultaneously established that flight control could be maintained, even when both halves of the brain were asleep. Perhaps wisely the frigatebirds tended to be a little higher above the sea when asleep (160 m) than when awake (135 m).[8]

The tactics used by travelling penguins as they negotiate waves, and by flying seabirds as they wrest energy from air moving above sea, do not bring the story to an end. These tactics are employed at relatively small spatial scales. Let us extend the story up the spatial scale.

* * *

Particularly among those birds that roll from side to side as they maintain an intimate relationship with the wind – one might almost say kiss the wind on their journeys – the journeys are by no means straight. Albatrosses employing dynamic soaring are zigzagging at the scale of 100 metres or thereabouts, and this pattern has been picked up by GPS tracking.[9] Similarly with shearwaters, and a predictable outcome is that the speed with which the birds travel between two fixed points is lower than expected, simply because they are travelling further than the straight-line distance. Exactly this has been observed by Tim Guilford and colleagues tracking Manx Shearwaters between Skomer and their feeding areas further north in the Irish Sea. The speed of the shearwaters' directional travel was about 40 km/h, rather less than 50 km/h, which is the speed at which Manx Shearwaters are calculated to cover most ground (or, more precisely, salty water) per unit of energy expended during powered flight. The discrepancy is very probably due to the shearwaters' use of slope and/or dynamic soaring, and hence a zigzag track, to reduce the cost of their journey.[10]

This increase in the tortuosity of the track is likely to be greatest when the bird is travelling across the wind.* When the wind is astern, the bird can spread its wings and celebrate on the downwind rollercoaster. When it is necessary to travel into wind, harder work and a substantially increased heart rate are in prospect. They are unwelcome and best avoided. No wonder the full circuits of the Southern Ocean made by albatrosses are from west to east, driven downwind by the westerlies. Illustrating the same point, Cory's Shearwaters nesting on the Berlengas islands off Portugal mostly headed north-west or southwest when departing the colony to collect food for their chicks. Since the prevailing winds during this September 2006 project were from the north-east, the birds apparently favoured crosswinds or tail winds, and avoided journeys into wind.[11] But perhaps this study needs viewing through the spectacles of scientific caution since, had the birds headed north-east into wind, they might have crashed into the coast of mainland Portugal barely 20 kilometres from the Berlengas!

While zigzag seabird tracks can be a consequence of energy-saving exploitation of the wind, there is another possibility. Animals spanning the range of complexity from the simplest worms to magnificent mammals show a behaviour known as area-restricted search. On encountering food, they tend to turn more frequently than previously. This keeps

* Even if tortuous zigzagging tracks create complications, the difference between a bird's airspeed, known fairly accurately, and its ground speed, measured by attached GPS devices, could provide a measure of wind speed and direction at the bird's location. The possibility of using this information to supplement that from weather buoys and so improve weather forecasting is now being discussed. http://www.pnas.org/content/113/32/9039.abstract (accessed 14 June 2017).

them in the area where the food was encountered, potentially increasing the chance of another encounter, especially if their food tends to occur in clumps. And of course seabird food, be it a shoal of fish or a krill swarm, tends to occur in clumps. Therefore, if the nematode worm *Caenorhabditis elegans* with a mere 1,000 body cells can include area-restricted search in its lifestyle, there is every likelihood that seabirds also can. Indeed they do. The way scientists use this variation in behaviour to identify areas where tracked seabirds are finding food will be discussed in Chapter 8. For now, the point is that the behaviour will generate more convoluted tracks over the scale of hundreds of metres or a few kilometres. In Wandering Albatrosses, this zigzagging associated with area-restricted search occurs at a scale of about one kilometre – as compared to the 100 metre scale of dynamic soaring – necessarily modifying the relationship between the albatross's track and the wind direction.[12] This could have energetic consequences, as has been demonstrated in Northern Gannets.

Courtesy of a research team led by Françoise Amélineau,[13] French gannets from the colony on Île Rouzic, off Brittany, bore a trio of devices – accelerometers, GPS and depth recorders – when setting forth to collect food for their chicks. When these devices were retrieved, Amélineau could calculate the amount of energy the gannets were dissipating at various stages of their journeys from Rouzic, the sole French gannetry. It transpired that the straightline journeys to and from the feeding areas off south-west England were less costly, in terms of calories expended per minute, than the spells when the birds were twisting and turning above fish below. Prior to this study, the extra costs of twisting turning flight had been predicted from aerodynamic theory, but the gannet work was a pioneering direct demonstration of the extra cost of tortuosity.

* * *

Expanding the spatial scale once more, a breeding seabird is making a succession of journeys back and forth to wherever it feeds. If these journeys are modest, say up to 100 kilometres, then the potential for adjusting the route between colony and the food to exploit large-scale wind patterns is slight. That potential expands when the bird is making loops of thousands of kilometres from colony to food and back to colony.

In Chapter 5, I made mention of the 15,000 km journeys undertaken by Murphy's Petrels during spells of about 19–20 days when they are excused incubation duties because their mate is attending to the egg. These travels can take the birds over 4,000 km north-east from Henderson Island, in the bluest, emptiest expanses of the South Pacific, towards South America. Remarkably, our GPS tracking showed how, during the first day of the journey, the petrels head approximately southward from Henderson (24°S) to latitudes around 35–40°S. Once that far south, they start to turn left towards the east, benefitting from the west winds of these higher latitudes. The pattern is set for an anti-clockwise loop that extends towards South America, and then continues during the homeward journey. (See Map 7.) When compared to the outbound leg, returning birds follow a more northern track (15–20°S), relishing the helping impetus of the south-east trades of these lower latitudes.

Not all Southern Hemisphere loops are anti-clockwise like those of Murphy's Petrels of both sexes. In other species, clear route differences between males and females have been discovered. While an entire book was required to describe how men hail from Mars and women from Venus, I can report on the clockwise tendencies of male Wandering Albatrosses and the anti-clockwise predilections of females in a single paragraph.

The Wanderers nesting on the Crozets (46°S) in the Southern Ocean find themselves in one of the windiest parts of the world. Fiercely the west wind blows day after stormy day. When embarking on foraging trips during breeding, the albatrosses normally leave their colonies in a downwind direction, towards the east.[14] Journeys that go north of the archipelago, most commonly undertaken by breeding females, are then anti-clockwise, while those to the south, habitually made by males, are clockwise. These strategies, respectively involving return journeys around 30 and 55°S, mean that the birds return to the longitude of the Crozets at latitudes where the easterly component of the wind is stronger than in the Roaring Forties. There are, however, some shorter journeys, for example during the chick-guard phase, when birds remain close to Crozet latitudes. They necessarily face headwinds on either the outward or homeward journey, during which they tack like sailing ships.

While the Southern Ocean is certainly dominated by westerly winds pushing a bird east, there are periods, time slots, when movement to the west is potentially easier. One such period is after the passage of a low pressure system. As the cold front trailing the system passes an island, so the wind swings quite abruptly from north-west to south-west. Eric Woehler, a Tasmania-based biologist with some 30 years of experience of southern hemisphere seabirds, tells me that this is when prions and diving petrels leave Marion Island to embark on a foraging trip. Similarly, it is in the aftermath of the cold front that Wandering Albatrosses in the southern Indian Ocean seize the opportunity to make progress westward, flying north-west at right angles to the south-west wind.[15] This takes them north of any high pressure system following in the wake of the low. There the wind is easterly, allowing further progress westward.

Although these studies show how breeding seabirds exploit prevailing wind patterns to reduce travel costs, they shed no light on how the birds would fare were wind patterns to change. We might reasonably guess that albatrosses could become becalmed, were the wind to fade away. But consider the alternative. Since albatrosses need winds of at least 30 km/h to allow dynamic soaring, might not winds stronger than those currently prevailing reduce their travel costs and enhance breeding success?

This possibility has been given credence by the hugely-informative long-term French study of Wandering Albatrosses on the Crozets.[16] Over the study's 40 years, winds have strengthened in this part of the Southern Ocean, both towards the north where breeding females concentrate their feeding, and further south where males feed (see Chapter 3). Stronger winds have allowed the birds to travel faster. Since their feeding success depends principally on distance covered, faster travel allows the albatrosses to gather the same quantity of food, mostly squid, in less time and arrive back at the nest more rapidly. This is especially advantageous during incubation since a major cause of breeding failure is abandonment of the egg by the sitting parent, urgently awaiting the return of the mate from its squid hunt. Associated with the improved breeding success and stronger winds, the birds have become about one kilogram heavier in the 20 years, 1990 to 2010,* without any increase in the lengths of their bones. Whether the weight increase is due to better, plumper body condition, or to the stronger winds allowing dynamic soaring by heavier – and hence faster – albatrosses, remains unresolved. Equally uncertain is whether these favourable circumstances will persist as wind patterns continue to alter under climate change.

* * *

Not only do seabirds adjust their routes in response to prevailing wind patterns when foraging from the colony, they also do so on a global scale, during migration. That is the cue for a brief sketch of global wind patterns, due in part to the Coriolis effect which is a result of the earth's rotation and the alleged cause of the proverbial tendency of the water draining down the bath's plughole to rotate clockwise in the northern hemisphere and anti-clockwise in the southern. Thus in the northern hemisphere westerly winds dominate across the North Pacific and Atlantic at latitudes of 40–50°N. Further south, at the latitude of the Tropic of Cancer (23.5°N), the air is moving back east to west, the North-east Trades. Combine the westerlies to the north and the North-east Trades, and a clockwise circulation becomes established in the Northern Hemisphere.

In contrast, in the Southern Hemisphere the westerlies blow most steadily at 40–50°S, with barely any land except Cape Horn to interrupt their flow. But the return flow, east to west, is here the South-east Trades. Combine the westerlies to the south and the South-east Trades, and an anti-clockwise circulation becomes established in the southern hemisphere.

These patterns were well-known to mariners in the nineteenth century and reflected in the routes taken by wool clippers plying between England and Australia or New Zealand. Leaving England, the route was southerly as far as the Canaries or Cape Verdes, off the Sahara. At that point, the North-east Trades were encountered. The ships headed west

* In 1990 females weighed about 8 kg and males 9.5 kg.

of south, crossing the Equator in the direction of Brazil, all the while hoping not to become becalmed in the calm winds of the Doldrums or hindered by equatorial westerlies. Once well south of the Equator, the clippers gradually turned to the east. On reaching latitudes south of the Cape of Good Hope at the southern tip of Africa, they could hope for a brisk downwind run eastward to Australia or New Zealand.

Returning laden with wool from the Antipodes, the ships again benefitted from the westerlies, all the way across the southern emptiness of the Pacific. Once Cape Horn, at the southern tip of South America, had been rounded, they began to head north in the South Atlantic. However, to exploit the westerlies and then the South-east Trades, the northbound track in the South Atlantic was commonly to the *east* of the southbound track. That southbound track was crossed around the Equator where again the Doldrums, sandwiched between the two trade wind zones, posed a possible obstacle. Then, in the North Atlantic, the ships were *west* of their southbound track, hoping to catch the westerlies to provide a final spurt to England and home. The outcome was a figure-of-8 track in the North and South Atlantic, reflecting the large-scale wind pattern across northern and southern hemispheres. At the broad scale, the same pattern prevails in the North and South Pacific. It is a pattern that influences the migrations of seabirds, and is indeed reflected remarkably closely in the tracks of some species.

Exactly this figure-of-8 track was described in Chapter 4 for those Manx Shearwaters travelling between their Welsh nesting stations and the non-breeding seas used off the coasts of Brazil and Argentina. Another species making trans-equatorial migrations is the Sooty Shearwater. Those breeding in the New Zealand region head to the North Pacific for the austral winter while those breeding in the Falklands go to the North Atlantic. One important colony in the Falklands is the 33 hectare kidney-shaped Kidney Island. It is home to over 100,000 breeding shearwaters that April Hedd of Newfoundland's Memorial University was brave enough to study. I admire her bravery not so much for risking scars from the shearwaters' sharp bills but for risking encounters with South American Sea Lions. Shearwaters choose to site their nesting burrows amid clumps of tussac grass as high as a person.

Sea Lions like to rest in this terrain, and I know from experience how easy it is to round a tussock on Kidney Island and meet a surprised 200 kg sea lion bull with bad breath and a grumpy demeanour.

Fortunately Hedd survived and was able to report the geolocator tracks of 19 shearwaters recorded over one or two years.[17] (See Map 8.) These tracks perfectly show figure-of-8 return journeys between the Falklands and the principal non-breeding area, the continental shelf waters brimming with Capelin off the eastern Canadian Grand Bank. Most starkly, there was virtually no overlap between northbound and southbound tracks in the North Atlantic, respectively west and east of the Mid-Atlantic Ridge. That, of course, helps explain why European birdwatchers are far more likely to see Sooty Shearwaters in the northern autumn when the birds are heading back to their Southern Hemisphere colonies, mostly along a surprisingly narrow corridor.

The shearwaters' S-shaped sigmoid northward route is remarkably similar to the northward journey described in Chapter 4 made by Arctic Terns heading to breed in Iceland and Greenland.[18] Both species are exploiting the same winds at roughly the same time of year. Perhaps what still requires explanation is why a minority of species, such as Sabine's Gull, apparently ignore the potential advantages of a following wind.

The figure-of-8 trans-equatorial journeys in the Atlantic are repeated with quite remarkable fidelity in the Pacific, for example by Short-tailed Shearwaters.[19] After breeding on islands off Tasmania, most tracked birds venture well into the Tasman Sea before heading north to two critical non-breeding areas of high productivity in the North Pacific, one off Hokkaido, Japan, the other surrounding the Aleutian Islands stretching outwards from mainland Alaska. All the tracked shearwaters begin their return migration to the breeding colony between mid-September and early October. Birds travel in a south-westerly direction through the central Pacific, skirting west of the Hawaiian Islands. After crossing the Equator, the shearwaters continue south-west until reaching the east coast of Australia. They then follow that coast southwards, passing the subtropical surfing hotspots of the Gold Coast before making landfall on the bleaker Tasmanian nesting islands.

One potential obstacle to these figure-of-8 routes is the Equator. Hereabouts there may be an absence of wind, doubtless a significant expla-

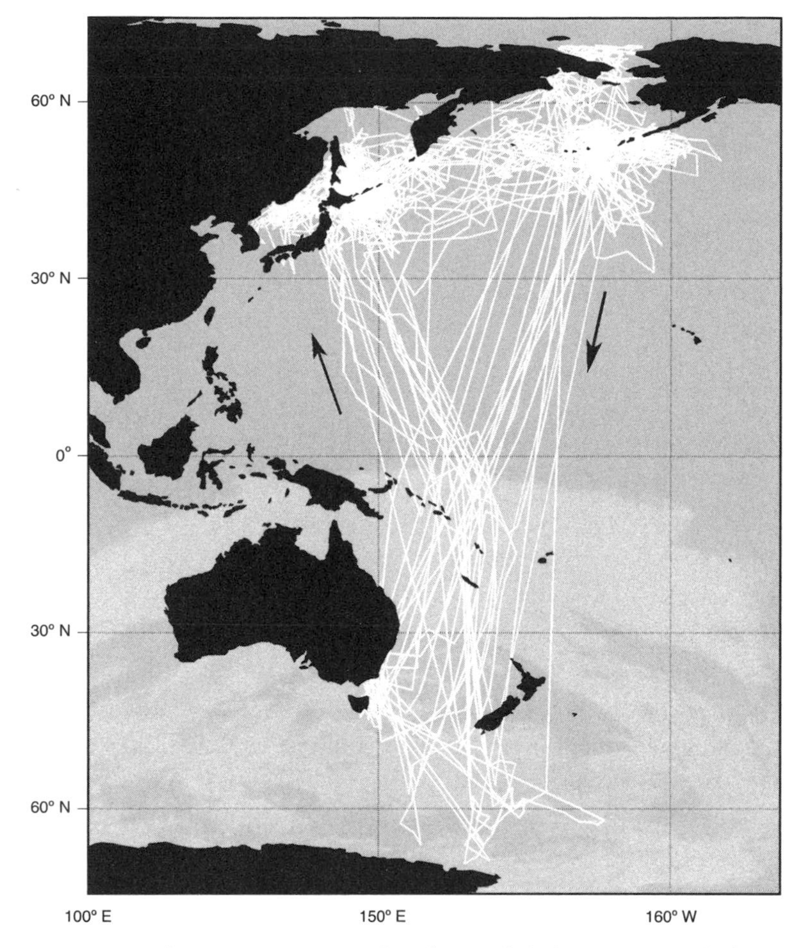


Figure-of-8 migration routes of 16 Short-tailed Shearwaters tracked from Great Dog Island, Tasmania. Reproduced with permission of BirdLife Australia, from the work cited in Note 19, Chapter 6.

nation for the near-absence of albatrosses from Equatorial regions.* Or there may be a band of westerly winds, blowing along the Inter-tropical Convergence, and clearly unwelcome to a southbound bird tracking south-west. Cory's Shearwaters, a species met in Chapter 4, breeding in the North Atlantic and occupying several distinct South Atlantic

* The one exception is the Waved Albatross, a species largely confined to nesting on the Galápagos and feeding off Peru.

non-breeding areas, might encounter this energy-sapping obstacle when heading south in the autumn. However the equatorial westerlies blow most reliably from June until the beginning of November, just before the mid-November passage of Cory's Shearwaters through the region.[20] Of course it might be a coincidence that the birds' passage through the area occurs just as the winds abate. Or it might not!

* * *

In the previous sections, I have attempted an overview of how many aspects of seabirds' lives are adapted to make efficient use of the winds that are such a feature of their marine domain. But the weather is variable. That is a statement of the obvious. The equally obvious corollary is that, at least sometimes, the weather can disrupt the birds' lives. The disruption can run the full gamut, from a storm, which is a troublesome but temporary disruption to provisioning the young, to a full-blown life-threatening hurricane. Now then is the moment to describe how modern studies have revealed how horrid weather has inconvenienced seabirds, or worse. This is particularly relevant in current times when climate change models predict more variable weather, and a greater likelihood of stormy weather, for many parts of the world.

At the inconvenient end of the range of impacts is the effect of wind on European Shags feeding off Scotland's east coast. Because the shags come ashore at night, the time spent foraging each day, as evidenced by daily number of hours with wet feet, could be determined when the immersion loggers on their feet were retrieved one or two years after first deployment. What Sue Lewis of Edinburgh University discovered was intriguing.[21] Daily foraging time was lower when the wind was in the west than in the east, the latter associated with a long fetch across the North Sea and hence rougher seas. However daily foraging time dropped as wind speed increased. Putting it anthropomorphically, it perhaps became a waste of time for the shags to search for food in seas stirred up by stronger winds. But the decline in foraging with wind speed was more pronounced in the larger males than the smaller females.[*] Just why

[*] Males are about 18% heavier than females.

the females were the more inclined to continue feeding in worsening conditions remained unresolved.

Just as wind can cause problems for seabirds, so can rain. A nice example comes, unexpectedly, from Antarctica, the earth's driest continent.* In late 2013, young downy Adélie Penguin chicks, beginning their lives in the continent's French sector, experienced temperatures five degrees Celsius above normal. Precipitation, normally snow, fell as rain which drenched the chicks and caused the death of about half of them.[22] Alas, the parents were not in a position to remain with the chicks to provide shelter. This was because the same weather that brought rain had failed to clear the ice close to shore. Thus, when foraging for their chicks, GPS-tracked parents were forced to travel twice the distance they covered in the previous season and were away at sea for around five days instead of three.

Without doubt, the weather can at times wreak greater havoc on seabirds. In particular, prolonged stormy weather can disrupt feeding opportunities. Weakened seabirds are then blown ashore, sometimes in huge numbers. In July 2011, the number of prions washed onto New Zealand's shores perhaps exceeded one quarter of a million.[23] Such large-scale strandings of seabirds are called 'wrecks'. Leading sometimes to birds being spotted far inland and in the oddest places, wrecks are by no means confined to the Southern Ocean. They can occur in the North Pacific affecting, for instance, Cassin's Auklets. A North Atlantic species that is particularly vulnerable is the Little Auk or Dovekie which, moving south in winter from its breeding strongholds of Svalbard and Greenland, is frequently wrecked both on the coasts of Europe and the eastern seaboard of North America, although not necessarily simultaneously. In their classic book *Seabirds*, published in 1954, the pioneer seabird ornithologists, James Fisher and Ronald Lockley, write of Little Auk wrecks. "When wrecks occur the light, small-winged little birds turn up in all sorts of places: on reservoirs, lakes, ponds, duck-ponds, rivers, sewage-farms, flooded gutters; in greenhouses, down chimneys, in

* Annual precipitation along the coast of Antarctica is typically around 200 mm, a little more than one-third of the annual total outside my Cambridge (UK) window, looking out onto one of the driest regions of England.

porches, back yards, pigsties, gardens, roads, turnip-fields; and are caught by foxes, dogs, opossums, raccoons, gulls, ravens, crows and boys. During the 1911–12 wreck of the Little Auk a doctor in Finsbury* met one entering his surgery door; it snapped at anyone who tried to handle it."[24]

Wrecks are not solely a feature of higher latitudes. They may also occur at lower latitudes, especially in association with hurricanes. For example, in 2008 hurricane Gustav brought wholly marine species, including Magnificent Frigatebirds and Sooty Terns, to the land-locked US state of Arkansas.[25] Those stray birds provided a thrill for bird-watchers but, arguably, their straying, and likely deaths, had little impact on the worldwide populations of frigatebirds and terns.

Nonetheless there are occasions when hurricanes exert extreme impacts on seabird colonies. Bermuda holds modest numbers of Common Terns nesting 1,000 km from the nearest North American colony. In early September 2003, when the hurricane Fabian passed over Ian Nisbet's small study colony of 10–30 pairs, all the breeding males were eliminated.[26] Only females survived to return in 2004. In the absence of a choice, the females paired with each other and inevitably laid infertile eggs. Over the subsequent years, males returned and chicks were reared once more. The guess is that these males were immature at the time of Fabian's onslaught, away at sea, and not in the vicinity of the colony. They survived to become fathers in later years.

A wise seabird might do well to avoid hurricanes and doubtless migratory routes have, at least in part, evolved to reduce the likelihood that birds will encounter severe weather. This is akin to yachtsmen habitually avoiding the Caribbean during the hurricane season. More intriguing is the possibility that birds can detect the approach of extreme weather from afar, and then take preparatory avoiding action. Exactly this ability appears to be within the skill set of the Golden-winged Warbler, a mite weighing about 9 g and one of the smallest birds to be tracked by geolocators.[27] In the second half of April 2014, five male warblers arrived back on the Tennessee breeding grounds. But trouble was brewing in the form of an approaching tornado which, across the central/

* A district of central London, not to be confused with Finsbury Park, the Underground station in north London where Arsenal supporters alight to watch football matches.

eastern United States, caused over one billion dollars' worth of damage and at least 35 human deaths. At least 24 hours before the storm's arrival in their territories, the warblers initiated an emergency evacuation, embarking on a 1,500 km clockwise circuit towards Georgia and Florida. After a five-day absence and the passage of the storm, the warblers returned to Tennessee to defend their breeding territories.

This extraordinary escape behaviour begs the question of what alerted the warblers to the approaching mayhem. One possibility is that infrasound, sound at frequencies below the range of human hearing,* travelled outward from the tornado storms with little attenuation, and was detected by the birds. Many bird species are known to be sensitive to infrasound.

If small warblers have the capacity to detect approaching storms, possibly via infrasound, possibly via changes in barometric pressure, it would be a surprise to me if seabirds do not share the ability. This awaits proof, but the 2nd World Seabird Conference held in the shadow of Table Mountain learnt how, when the Category 5 tropical cyclone Pam was rampaging in the south-west Pacific close to New Caledonia in March 2015, frigatebirds changed their behaviour. Normally remaining within 200 km of their colony at this season, they went 700 km away when the storm threatened. Did they get wind of the approaching cyclone?

Doubtless there will be occasions when avoidance fails, and seabirds become embroiled in an oceanic maelstrom. We do not know details of how they fare but the remarkable passage of a Whimbrel through a tropical storm gives cause for optimism. The Whimbrel was a contributor to a long-term study by Bryan Watts of Virginia's College of William and Mary. Using satellite tags, Watts has studied the Whimbrels nesting around Hudson Bay and points west.[28] After breeding, the Whimbrels face a journey to Brazil, following the eastern seaboard of the United States and then crossing the Caribbean at a time of year when there is a risk of encountering violent storms or hurricanes. In August 2011, one Whimbrel named Hope encountered Tropical Storm Gert off Nova Scotia. Undeterred, Hope entered the storm at a ground speed of about 12 km/h. Passing the eye, Hope emerged - or was ejected - 27 hours later

* Infrasound has a frequency below about 20 Hz or cycles per second.

at a speed of 150 km/h. Watts comments "She immediately changed course and took plan B, an escape route to Cape Cod [on the Massachusetts coast]. These birds have repeatedly shown some amazing situational awareness." If a shorebird the size of a smaller seabird can overcome such a storm, surely seabirds also can.

And that is the persistent message emerging as the interaction between seabirds and their windy domain is studied. Seabirds are able not merely to cope with an environment that, to landlubbers, seems so alien. They can thrive when flying over or swimming through the waves. Of course, mere travel is insufficient. The birds must find and catch food, as we shall shortly explore.

## CHAPTER 7

# Stick or Twist?

## The Consistent Habits of Individuals

Many animals are creatures of habit. Humans certainly are. Some of these habits are supremely trivial. I always begin shaving with a downward stroke across my left cheek. Other habits are more significant, indeed potentially life-saving. Where road traffic keeps to the left, a parent will teach a child to look right, then look left and then look right again before crossing the road. And some habits perhaps reflect more subtle interactions between people at large. We have all had the experience of explaining our route between two places, perhaps between home and the workplace, only to be amazed to discover that someone else, making virtually the same journey, takes a wholly different route.

This is not as stupid as it might seem. Perhaps the fact that the two travellers, or indeed travellers more generally, choose a variety of routes means that no one route becomes clogged with traffic. Everyone benefits from the multiple route choices that have been made.

I have been lucky enough to study Manx Shearwaters over many years. Unfortunately for the birds, the studies have meant that they are unceremoniously removed from their burrows time and again. Some are stoical, some more rumbustious and aggressive. There are clear differences in character between the birds. And, as has emerged in other comparable seabird studies, a minority of pairs raise the majority of chicks. This variation in quality bears comparison with the so-called Pareto Principle which originated in the observation that 80 percent of Italy's wealth belonged to only 20 percent of the population. It has since been extended to many aspects of organisations. For example, 20 percent of the workers produce 80 percent of the results.*

With individual seabirds probably differing in quality and character and certainly living long enough to develop individually consistent habits, it is perhaps no surprise that individual consistency is a theme that emerged once seabird tracking projects proliferated. This chapter will describe examples of such consistency, now identified in over 10 percent of all seabird species, and mull over whether it benefits the birds.

The consistency might operate over several timescales. An adult bird might set forth from the colony to find food in a direction that is more consistent than the various directions chosen by fellow colony members. That is a short-term choice. On a longer timescale, habits acquired as a youngster might be retained into adulthood. Over an entire lifetime, migrants might return to the same non-breeding area year after year.

The consistent use of certain areas is conceptually distinct from other possible manifestations of consistency. Some birds might regularly choose to search for fish in cooler waters, others in warmer waters. Some might choose to follow ships while others are more wary. In practice these alternative types of consistency may not be completely distinct

* The principle also appears to apply to writing seabird books; about 80 percent of the useful words come in 20 percent of the screen time, while the remaining 80 percent of time mysteriously vanishes.

from spatial consistency; for example a chosen area may be characterised by warmer water than used by most other members of the species.

Birdwatchers will be familiar with lines of Northern Gannets heading in a chevron formation to who-knows-where. In fact, individual gannets leaving the breeding colony in search of food tend to depart towards the feeding grounds in a flight that is straight and consistent. At least that is true for the gannets setting forth from Bass Rock in the outer Firth of Forth, Scotland. They have two preferred directions, one north-east to an area north-east of Aberdeen, and another south-east to seas off north-east England. All nine birds satellite-tracked by Keith Hamer, then of Durham University, showed a decided preference for one or other direction, and indeed four 'south-east' birds shunned the north-east alternative altogether, at least when carrying a tag.[1] Only when gannets are in the general area where food is likely do they adopt a more tortuous searching flight path, so-called area-restricted search. Doubtless they are keeping a lookout for other fishing gannets, a surefire signal of food.

The consistent departure direction admits of two explanations. The first is that gannets do indeed have personally preferred foraging directions. An alternative is that gannets return to an area where they have found food quickly, only switching to another area if the search has been forlorn. If birds were mostly following such a 'Win/Stay: Lose/Shift' strategy, the colony observation would be of most birds consistently heading out in their individually chosen direction - but sometimes switching. And switching would be more likely the longer the observation period. Ewan Wakefield returned to the Bass Rock around a decade after Keith Hamer's study. Following gannets over two weeks, time enough for seven trips, Wakefield reported no decay in consistency over time.[2] The gannets were also consistent from one year to the next. They really did seem reluctant to change their spots and did not employ the 'Win/Stay: Lose/Shift' strategy.

Not only do certain gannets prefer certain directions, there are gannets with distinct feeding preferences. More precisely, there are some that routinely scavenge behind fishing boats, and others that virtually shun this type of feeding. The gannets in question were studied by Samantha Patrick at the Welsh colony of Grassholm,[3] a small lump of an island that, from afar, looks white. Only when nearing the island does one realise that the white coating is not an accumulation of droppings but a noisy living carpet of tens of thousands of adult gannets. But the study birds did not simply carry GPS loggers, seven also bore time-depth recorders. Thus Patrick's team could identify when and where, on their foraging trips, the gannets were diving. Couple this information with high-resolution data on the location of commercial fishing vessels, now available for vessels greater than 15 m fishing in UK waters (regardless of nationality), and it was possible to discover whether every gannet dive was close to a vessel or not. With around 200 dives available for each gannet, it transpired there were two birds that nearly always* dived near a fishing vessel, three that showed flexibility, and two that never dived near a vessel. However there was no sign that foraging effort, measured for example by trip duration, differed between scavengers and non-scavengers. Nor was there a difference in the body condition of the birds. It looked as if the two habits delivered equivalent benefits, a point to which we shall return.

Such consistent patterns are not the sole preserve of flying seabirds. At least some penguins follow suit. King Penguins in the Falklands, feeding large chicks during the southern winter, consistently head north.[4] Their trips, reaching a distance of about 300 km from the study colony at Volunteer Point, mostly target the margin of the Patagonian Shelf where the shallow shelf (water about 200 m deep) begins to slope downward to the 5,000 m abyss of the South Atlantic. But, as often with scientific studies, it is right to strike a cautionary note. The northward outward journey meant the penguins benefitted from the north-flowing Falklands Current. Perhaps the consistent penguins all had the wit to recognize and utilise a watery helping hand.

Modern tracking has frequently identified repeated use of particular areas by individual breeding birds. That opens the possibility that migratory routes of individual birds might be consistent and consistently different from those of their fellows. That is certainly true of Wandering Albatrosses on the French Iles Crozet and Kerguelen in the Southern Ocean. Wonderful studies have tracked adult albatrosses in the interval

* At least 80 percent of dives were near a vessel.

between one breeding attempt and the next. In a nutshell, the albatrosses occupy those spare months with one of four travel patterns; stay in local waters in the Southern Ocean, stay locally but occasionally venture downwind towards Australia, migrate as far east as New Zealand, or embark on a grand circumpolar tour. Amazingly the Crozet birds mostly adopt the stay-near-home option while the Kerguelen birds are the Wanderers that truly wander on circumpolar journeys. Incidentally, since the birds on the two archipelagos are known to be genetically alike, this argues against consistent patterns being the result of genetic inheritance. And these travel patterns are consistent. Once a bird from either archipelago has reached adulthood, it will persist with the same migratory habit, sedentary or migratory, for many years.[5]

Migratory consistency extends to the Atlantic and doubtless beyond. Few visitors to the subtropical jewel that is Madeira will take much note of the Ilhas Desertas, a barren string of islets to the east of the principal island. One of the islets, Bugio, is home to a very rare gadfly petrel, the Desertas Petrel, with a world population of just 200 breeding pairs. When not breeding in winter, each petrel heads to one, and only one, of five distinct areas across the ocean. Two of the areas are north of the Equator, in the Gulf Stream off the south-eastern United States, and around the Cape Verdes. Three are south of the Equator, off northern and southern Brazil, and in the central South Atlantic. When 26 birds were tracked over two or three winter periods, every one remained faithful to its chosen ocean region.[6]

Not only is there evidence of fidelity to migratory routes, there is also evidence that the timing of migration can be characteristic of the individual. For example, the Black-browed Albatrosses nesting on tussac-clad Bird Island, South Georgia, have been a source of inspiring information. When not facing breeding responsibilities, most individuals head eastward to the Benguela Current off South Africa. Their sojourn there is characterised by remarkable individual consistency in timing. For example, at the end of the winter, the dates at which individuals leave south-west Africa, arrive in the south-west Atlantic, and arrive near to Bird Island are strongly correlated from one year to the next. One bird bucked the geographical trend, wintering off Australia instead of Africa, but it did not buck the timing trend. Its dates of departure from

Australia and arrival in the south-west Atlantic were 10 August and 13 October in 2002, and 12 August and 9 October in 2003![7] Sixty-four days to beat into wind halfway round the world one year, a mere 58 days the next.

Migration takes birds to non-breeding areas, most commonly in winter – and here again evidence for consistent choices emerges. Some of the evidence comes from traditional hard work and, more specifically, from scouring the east coast of Scotland on bleak winter days for European Shags bearing colour rings applied during the summer at the colony on the Isle of May. Well wrapped up, Hannah Grist's team accumulated 3,797 winter resightings of 882 individuals, up to 486 km north and 136 km south of the colony.[8] Individual Shags tended to be sighted the same distance from the colony, not only in the short term, when sightings were mere days apart, but from one winter to the next. They evidently developed an affiliation for a stretch of coastline.

To discover whether offshore species are also faithful to wintering areas, modern devices such as geolocators are needed. On Skomer Island, Wales, adult Atlantic Puffins have delighted the island's many day-visitors for decades. Only recently have the winter movements of the Puffins begun to become clearer, thanks to the research of the OxNav group, presaged in Chapter 4. In winter Skomer Puffins essentially make one of four possible migratory journeys: remain in local, north-east Atlantic waters,* remain in local waters and then head into the Mediterranean before returning to Wales, cross the Atlantic and then loop northwards to Icelandic waters before the return, and reach the central Atlantic before entering the Mediterranean in advance of the return to Wales. (See Map 9.) When the Oxford group had accumulated up to six years of geolocator data for some birds, it was striking how individuals were consistent from year to year despite the diversity among the whole population.[9] Such a variety of complex routes seems unlikely to be programmed genetically, nor, remembering that the baby 'puffling' fledges unaccompanied at night, is it likely that the route is learnt by following parents.

* I am obviously using 'local' quite loosely!

The most recent research on the Skomer Puffins has hinted that the birds that reach the Mediterranean in late winter – either after remaining in local waters or after venturing into the more distant Atlantic – enjoy the highest breeding success the following summer. Acknowledging that flying taxes puffins, one interpretation is that only higher quality birds, perhaps those with superior fishing skills and consequently better body condition, can reach the distant Mediterranean. Once there, they can benefit from a late winter spurt in marine productivity. Returning to Skomer replenished, these high quality birds are more likely to proceed to rear a chick than their lower-quality neighbours.[10]

While consistent habits are certainly a feature of seabirds' lives, their prominence can vary over time. For example a team under the aegis of the Ascension Island Government studied frigatebirds on that island in the tropical Atlantic, and amassed data from a mighty impressive 776 Ascension Frigatebird trips. Compared to trips in the hot season in the early calendar months of the year, trips during the cool season, the latter calendar months, took the birds further from the colony, covered more kilometres and lasted longer. They also showed less variation. Perhaps feeding conditions were more benign during the hot season, allowing the birds various options – hence variable trips.

Most consistency studies I have described concern adult birds. This is simply for reasons of practicality; it is far easier to repeatedly handle an adult bird at a colony than a youngster cruising the world's oceans during its years of immaturity. But this raises a very obvious question. Do young birds leave the colony with their own personal set of quirks, or are the early years spent trying various possibilities, in the manner of a teenager causing angst for parents, before settling into the relative rut of adulthood?

This is taxing to answer because of the difficulty of following birds in detail through those 'lost years' of immaturity (Chapter 3). However, as I write, there is no solid evidence that preference for particular foraging sites is inherited, in contrast to preference for migratory directions, which is almost certainly partly inherited from the bird's parents in many species. Another possibility is that the birds lacking consistent habits are those more likely to die in their early years. If this was the case, the survivors, the adults, would show more consistent habits than younger birds. Again there is no evidence for this scenario. Thus the front-running idea is that consistent habits develop over a bird's early years – and the limited evidence supports this.

Speaking at the 2nd World Seabird Conference, Steve Votier, a key contributor to the Grassholm studies, described how gannets with chicks undertook repeatable stereotyped journeys out of the colony. On the other hand, immature birds, followed by satellite-tracking, showed highly variable foraging locations and low route fidelity from one trip to the next. There was next to no individual specialisation. This difference was not simply associated with having chicks to rear. Failed breeders also showed considerable trip consistency (albeit less than breeding adults with chicks). Consequently Votier concluded that consistent individual foraging habits are learned during seabirds' long lives.

When in life is this consistency acquired? This remains a tricky question because fledging birds may be tracked for the first year or two at sea. Then there is a gap in knowledge, around the period when birds are 3–4 years of age and, literally, beyond the grasp of researchers. These difficulties notwithstanding, Tommy Clay has tackled the issue, using the exceptional data available on the Wandering Albatrosses of South Georgia, thanks to many years of effort by the British Antarctic Survey. Some birds develop consistency in the sea areas utilised in the first two years of life. Others take several years, but still achieve this characteristic before they begin to breed. But this variation has few obvious repercussions. Clay found little association between when a Wanderer developed consistent use of particular sea areas and either its likelihood of joining the breeding population or the age at which it would do so.

If consistency develops during an individual's lifetime, does that imply that a consistent lifestyle brings detectable benefits? Actually it might not. Suppose, very simply, that there was ten times as much food available in area A as in area B. Then if ten times as many birds visited A as visited B, the birds visiting the two areas would prosper equally, and a bird consistently visiting A would fare as well as one visiting B. A researcher would struggle to detect any advantages arising from the birds' consistency. Perhaps as a result of learning, the birds in my simple example have chosen the areas according to what biologists term the ideal free distribution, akin to the choice of various routes between home and workplace that was mentioned at the chapter's outset.

In an ever-changing world, there can be no assurance that the relative merits of areas A and B are forever fixed. One may improve, the other deteriorate. Putting it anthropomorphically, the bird does not want its fixed routine to cause it to miss out on a bonanza elsewhere, or to be stranded in an area where fish stocks have plummeted or the sea unexpectedly frozen over. There may be advantages in occasional sampling visits to unfamiliar areas. Possibly this happens. Studies of consistency over several years typically report lower consistency values than those conducted over a couple of years. Even among shorter studies, there can be a divergence between birds showing striking consistency and those, the putative samplers, showing none at all. For example, after leaving colonies in eastern Canada, Brünnich's Guillemots head into the Labrador Sea for the winter. Most birds spent successive winters fairly close to where they spent the previous winter, but a few clearly did not.[11] (See Map 11.) This was manifest in the distance between the birds' centre of activity in successive winters; it ranged from 22 km, implying these birds were bobbing on precisely the same seas from one winter to the next, to 1,200 km, implying use of a sea area half an ocean away. Plausibly the same sampling argument might apply to the Cory's Shearwaters encountered in Chapter 4. Most are faithful to a wintering area, but a handful switch to an alternative, for no very obvious reason.

Although the idea of the ideal free distribution posits that the birds join one or other group at random, birds could join groups according to their gender. This would achieve the same outcome, dividing the species' feeding activities between two (or more) areas. We have already seen how, in the period before laying, male gadfly petrels tend to travel further from the colony than females. Partly because of their greater weight and ability to cope with stronger winds, male Wandering Albatrosses routinely feed to the south of females. And there are consistent, gender-based differences in the species that has been prominent in this chapter, the Northern Gannet.

Adult gannets weigh about three kilograms, and females are larger than males by some 200 g. Not only do the birds leaving the Bass Rock colony show consistency of departure direction. Compared to females, males are fonder of the north-east choice, making shorter trips (in distance and duration) and shallower dives in waters where there is tidal mixing and prey is to be found at modest depths.[12] Probably because they are heavier, the females tend to make deeper dives than males in waters further offshore where the water column is stratified and prey fish swim deeper. Thus the consistently different behaviour of males and females helps distribute the birds across the North Sea, and reduces the degree of mutual interference, to the benefit of both sexes.

'Practice makes perfect' is a familiar aphorism. It is plausible that simply by consistently using a sea area, a seabird gains familiarity and hence an advantage. It may know where to find food just as we can complete shopping more quickly in a familiar supermarket where we know which shelves carry such favourite items as peanut butter, ground coffee and sardines. And if that sounds far-fetched, we shall sketch evidence in the next chapter that feeding seabirds target such more or less permanent marine features as ocean fronts, the supermarket shelves of the sea.

Despite these several plausible benefits of consistency, demonstrating its value to the perpetrators – the birds themselves – remains extraordinarily difficult. There is an important logical reason why this should be so. For example, Atlantic Puffins from the Isle of May colony, off eastern Scotland, can be fitted with a geolocator device which, when retrieved, reveals whether its wearer has spent the winter in the North Sea or ventured into the North Atlantic Ocean.[13] This is very likely a consistent choice, just as is the wintering area of Skomer Puffins. But – and here's the rub – the only puffins that yield information are those that return to the Isle of May where their devices can be downloaded. There is zero information on the wintering areas and consistency (or otherwise) of those puffins that die at sea. Overcoming this problem will be tough technically. The researcher needs to be confident that the yet-to-be developed device is correctly reporting the whereabouts of its wearer at the time of its demise far far out to sea, and that the (lack of) signals do not represent battery failure or accidental detachment of the device.

If the link between consistency and bird's chances of survival is elusive, there are at least hints that consistent birds breed more successfully.

For example, 91 Black-browed Albatrosses were studied, via GPS backpacks, during the chick-guarding phase at Iles Kerguelen by Samantha Patrick and Henri Weimerskirch.[14] The birds that subsequently raised the chick to fledging generally showed less variation, in maximum range (distance from colony to the trip position furthest from the colony, the terminal point) and the latitude and longitude of that terminal point, than did the birds that failed. Consistency paid.

Yet I cannot help wondering if the search for an advantage in consistency may frequently prove as illusory as the quest for that pot of gold at the end of the rainbow. Drawing on the idea of the ideal free distribution, the birds faithful to area A may fare as well in the lifetime stakes as those faithful to area B. And those two groups may do neither better nor worse than a third group that adopts a jack-of-all-trades lifestyle, missing out on the putative benefits of developing familiarity with an often-visited sea area, but having the compensation of sometimes hitting temporary jackpots, seething shoals of catchable seafood. Let us now explore where that food is most likely to be discovered.

CHAPTER 8

# Where Seabirds Find Food

A penguin jumps into the foaming sea to gather food after a spell ashore on the Falklands, and swims hundreds of kilometres before returning. An albatross embarks on a loop of thousands of kilometres, "scour[ing] the ashes of the sea where Capricorn and Zero cross",* before relieving its mate waiting stoically ashore on the egg on South Georgia. During these trips, the birds feed. That much is a statement of the obvious. The question is where. Do they feed along the entire route whenever and wherever swarming krill or choice floating morsels are encountered? Or

* From *Genesis* by Geoffrey Hill, https://www.theparisreview.org/blog/2016/07/01/genesis/ (accessed 14 June 2017).

do they have outward journeys with little feeding, followed by a period of feasting on plenty, followed by a homeward return?

This chapter will delve into where seabirds actually feed, and why they choose those feeding places, if indeed they do choose. In fact implicit in much of the earlier chapters is a presumption that researchers can pinpoint where seabirds actually feed, as opposed to where they merely pass by. It is therefore timely to pause to mention the types of evidence used to identify feeding areas. Without robust knowledge of feeding areas, it is meaningless to ask questions about their characteristics. (See Plate 7 in colour insert.)

If the bird enters a region where its mind turns from travel to food, its path across the sea is likely to become more sinuous. This could be particularly true after the successful capture of prey. Then, it makes sense to continue searching in the same relatively small area; where there's one fish, there could well be another. Tracking data, be it from satellites or GPS devices or geolocators can help identify these favoured areas. Most simply, they are areas where the bird spends more time, as compared to areas through which it quickly passes. Such areas can be pinpointed.

But the details provided by satellite and GPS tracks also allow more sophisticated identification of such favoured areas. In these areas, the bird turns more, adopting area-restricted search. This is precisely analogous to someone turning more, in response to cries of 'You're getting warmer', as she or he homes in on the thimble in a game of Hunt-the-Thimble. The turning can be measured. So too can first-passage time, the time required for a bird or indeed any animal to cross a circle with a given radius. The more its route twists and turns, potentially revealing a search for food, the longer will be that time.

Another approach to using tracking data to identify feeding areas can be illustrated by shopping. Suppose, with a heavy heart, you spend Saturday morning driving into town, first to fill up with petrol, a brief pause, and then to complete the main weekly grocery shop, a longer interval in the supermarket, before returning home. The rate at which your straightline distance from home is changing will be at a minimum, in fact virtually zero, during the period in the supermarket. Similarly a bird's rate of movement away from or towards the colony may reach a minimum when it is in a prime feeding region. That minimum displacement rate can identify feeding areas.

While movement patterns gives powerful clues to likely feeding, other devices give more direct evidence. Depth gauges – of which more in Chapter 9 – are telling. Very evidently, a species that feeds by diving, say a Common Guillemot or Sooty Shearwater, is far more likely to be feeding when making repeated dives than when floating on the surface or flying. Combine the depth recorder with a position, and it is possible to make fair inferences about where the diving bird is feeding. If the basic depth information can be combined with accelerometer information, providing a picture of the twists and turns during an underwater chase, then the chance of identifying actual feeding episodes improves.

For species that are predominantly aerial, say albatrosses, the number of landings per hour and total time on the water, as detected by tiny leg-mounted saltwater sensors, can reveal where a bird feeds. The principle is straightforward, but the complications are considerable. For instance, it may not be obvious whether a night spent on the sea is a night of feasting or of snoozing.

Finally, as mentioned in Chapter 1, there are devices which record when a bird opens its mouth and by how far, and devices which record a drop in stomach temperature. This drop will occur when food, roughly at sea temperature and thus cooler than the internal body temperature of a warm-blooded bird, is swallowed.

* * *

It could be that a seabird's approach to food-finding is supremely crude, to traverse the ocean in the hope of encountering, or perhaps I should say blundering into, food. The evidence that certain areas are consistently used would argue against this. So too would observations of birds streaming away from colonies in formations that speak of order. Halfway between Norway's North Cape and Svalbard, I remember passing Bjørnøya. The island was barely visible in the raw Arctic mist. That poor visibility was no deterrent at all to the Guillemots, both Common and Brünnich's, that streamed past our ship in organised lines or chevron formation as they left the colony. Surely these birds knew where they were going. This prompts the thought that there is information available at a colony to help guide a bird towards food.

If birds know where they are going, they could each have their own individual preferences or, as mentioned in the previous chapter, they could be following a 'Win/Stay: Lose/Shift' strategy. If the latter, they face a quandary after losing, and failing to find food. To shift at random would appear a poor strategy; there is so much ocean to be searched. One means of improving the chances of success might be to follow birds that had been successful and were returning from the colony to an active feeding area that had proved profitable on their last foray. Using traditional observation, tests of this idea have not been especially convincing – but there have been some encouraging results. For example, breeding Common Guillemots often gather on the water at the foot of colonies. Working in Newfoundland, Alan Burger wondered whether these assemblages could be where those adults that have not been blessed with recent fishing success obtain knowledge by watching their returning fellow guillemots.[1] In 60 percent of departures, breeding adults splashed down within this assembly zone, and were more likely to do so if they had not recently delivered a meal, or had spent more than an hour at the colony. Meals were subsequently delivered to chicks after 69 percent of such splashdown departures, but after 82 percent of direct departures.*

Another intriguing study was undertaken on Isla Pescadores,† off the coast of Peru and set in the immensely productive waters of the Humboldt Current. The island is home to many thousands of Guanay Cormorants whose dried guano was once extensively harvested as fertilizer before the advent of petrochemical-based fertilizers. Henri Weimerskirch's team used GPS tags to follow the Cormorants on their feeding excursions and depth recorders to pinpoint where they dived and fed.[2] There was no correlation between a bird's bearing when returning to the island and the bearing of its next outward trip. That is evidence the birds did not return to the same feeding zone from one trip to the next,

* The difference is highly suggestive of better fishing success among directly-departing birds but, in the jargon, not quite statistically significant.

† Island of the Fishermen.

presumably because shoals of the main prey, Peruvian anchovy, were extremely ephemeral. But the birds were not left without information about where to search. When setting forth, they briefly joined a large offshore raft of Cormorants swimming on the Pacific. This raft, a perpetually changing cluster of around 300 birds, was not aligned to the wind, but it was remarkably well aligned to the unbroken columns of Cormorants returning from successfully feeding on large prey patches. Thus the alignment of the raft gave departing Cormorants useful information about the direction of the currently-active feeding hotspot. And the information was used; the departing birds flew off in the indicated direction.

Once out at sea, an obvious clue to food is the presence of other feeding birds. Showing off his blue feet in a dance of self-important self-advertisement, a male Blue-footed Booby beside the tourist path on the Galápagos hopes to attract a female. However, attraction, if not love, is guaranteed if he starts to dive offshore. A booby slanting into Galápagos coastal waters that often seem so shallow as to render the dive foolhardy will assuredly attract other boobies hoping to share the fish shoal. This is a commonplace observation. Just as vultures keep a watch for other vultures as a guide to the presence of carcases, so seabirds certainly watch the activity of other seabirds (and fishing vessels: Chapter 10) as a shortcut to finding food. The fact the plumage of so many seabirds is substantially white, and conspicuous against the sea's surface, certainly increases the chance that feeding birds will be spotted from afar.

In the tropics shoals of tuna drive smaller fish to the surface. The water boils as the frantic prey fish try to escape their underwater nemesis, but there is of course a catch. Small prey fish at or near the surface can be caught by birds, ever on the lookout for tell-tale signs of shimmering fish scales. It seems very probable that the birds use sight to spot the opportunity.

What is intriguing is how the species of birds that gather above tuna may differ according to marine productivity.* For example in the east-

* This can be assessed by the density of chlorophyll in the water and also the depth of the thermocline, that sharp temperature discontinuity between warm surface water floating on top of denser cooler water below. A deeper thermocline is indicative of little mixing of the surface waters. Consequently the surface waters become stripped of key nutrients and correspondingly unproductive.

ern Tropical Pacific, when productivity is high, flocks above tuna are dominated by boobies, with petrels scarce.[3] As productivity falls, the flocks are composed mostly of Juan Fernández Petrels and Wedge-tailed Shearwaters. At the lowest productivity, a species with very low flight costs, the Sooty Tern, forms the core of the flocks, which then include few petrels. This pattern, to which we shall return, makes sense if, at low productivity, the only species able to exist are those with the low flight costs that enable cheap transit between rare and scattered food patches. At higher productivity it becomes possible for larger species to balance their energy budgets, and their presence excludes the smaller species. This exclusion likely happens in two ways, either the boobies jostle aside the smaller petrels and terns by their sheer physical size, or the boobies dive to catch prey at depths out of reach of the more aerial species.

In addition to vision, there is another sense that helps seabirds find food. It is the sense of smell, and it is especially useful for petrels and albatrosses, birds where that part of the brain involved in smell, the olfactory bulb, is notably large. Exploiting this sensitivity, offshore birdwatching trips routinely use chumming to lure rarely-seen petrels into view. They approach the vessel, and the delighted birders aboard, by flying upwind into the odour trail. Not only can petrels pick up the fishy smell of chum (and, incidentally, the smell of their own species), but they can also detect the chemical dimethyl sulphide, which is produced by phytoplankton in response to grazing by zooplankton and potentially provides petrels with a clue to the whereabouts of prey.

If birds are using smell to find food items, they might fly across the wind in the hope of detecting an odour plume drifting downwind from the food. Having detected the plume, they would then turn into the wind and follow the odour trail to source, possibly zigzagging as they approached. When Wandering Albatrosses were fitted with GPS-loggers to record their precise position and stomach temperature sensors to record the moment when food at Southern Ocean sea temperature entered the stomach, it transpired that about half the food items reached by flight (as opposed to by swimming) were approached directly.[4] The other half were reached by an abrupt turn that was followed by a direct or a zigzagging approach to the item, exactly as would be expected if smell was a clue to the possibility of lunch. On average the turn towards the point of eventual prey capture occurred at a distance of 1,300 m if the final approach was linear, and at 2,400 m if the final approach was zigzagging. The maximum distance albatrosses followed odour trails was an amazing 5–6 km.

Yet these ploys reveal far from the full story of the birds' search for food. There is now overwhelming evidence that birds focus their feeding attentions on certain specific areas of the ocean. Obviously these tend to be areas where success is more likely, where food is concentrated. That concentration can arise in several ways. It can arise because certain marine features, say the underwater topography, divert water flows. If the flow is towards or at least within reach of the surface, there is a possibility of concentrating both planktonic organisms and the larger animals that eat the plankton, and are themselves seabird food. Or – and this alternative is by no means sharply delineated from the first – local marine conditions may promote productivity. The amount of chlorophyll, the energy-capturing pigment that is an indicator of plankton presence (and hence marine productivity), may be high, increasing the likelihood of a concentration of seabird food. It is for this reason that many studies have correlated seabird abundance at sea with chlorophyll concentrations. Finally seabird food may be concentrated temporarily by the underwater activity of fishes or whales and dolphins, providing a brief bonanza for seabirds.

Where freshwater flows into the sea, there is also an influx of nutrients, creating the conditions to foster plankton growth and, ultimately, to attract seabirds. It is for this reason that the outflow of the Columbia River, from the United States' north-west coast, attracts great numbers of Common Guillemots and Sooty Shearwaters. On the other side of the Pacific, Wedge-tailed Shearwaters, GPS-tracked by Fiona McDuie from the Heron Island colony on the Great Barrier Reef, may visit freshwater plumes, especially those of the Fitzroy River. Discharging along the Queensland coast, the plumes enhance chlorophyll and presumably also prey abundance.

It has long been known that the input of freshwater from melting sea ice, which is of course fresh, creates a highly productive zone along ice margins, sometimes far from land. These zones represent hotspots – if the term can be permitted in such an icy context – for all sorts of wildlife from whales to seals to seabirds. Little Auks, pocket battleships of the High Arctic, are famed for their use of such ice edges. That habit might put them at risk as ice retreats in the face of global warming but, at least in the short-term, the conclusion could be premature. The High Arctic (80°N) archipelago of Franz Josef Land, well north of Russia, is now virtually ice-free during the summer. The region's Little Auks have lost their sea-ice-associated prey. Fortunately the retreat of large coastal glaciers has released large volumes of melt water, generating a combination of cold and osmotic shock that stuns the zooplankton in front of the glaciers and, for now, creates a summer feast for Little Auks.[5] Whether this will persist is open to question.

Half a world away, other birds also use the ice edge. This might be anticipated for Antarctic specialists such as Emperor and Adélie Penguins. It is more surprising for the lower latitude King Penguins raising their tubby chicks, downy and the colour of café-au-lait, through the sub-Antarctic winter. Remarkably, satellite-tracked parent King Penguins from Iles Crozet (46°S) head south at this season to gather food, mostly lanternfish (also called myctophids), for the chicks. This journey takes them about 1,600 km from the colony and as far as the marginal ice skirting Antarctica, where lanternfish are most abundant.[6]

Another unexpected 'habitat' exploited by feeding seabirds is the sea above moraine banks, higgledy-piggledy piles of stones and boulders left as underwater deposits by long-gone glaciers. An intriguing example of this comes from South Georgia where moraine banks occur at depths of 400–500 m at the margins of the continental shelf, before the shelf break and the drop to the abyss of the South Atlantic. The study subjects were Black-browed Albatrosses from the Bird Island colony, studied by the British Antarctic Survey for several decades.[7] Fitted during chick-rearing with GPS-loggers and stomach temperature sensors to pinpoint where food was swallowed, the birds took in most beakfuls above moraines. Just how a pile, admittedly a very large pile, of underwater boulders can lead to more krill and fish being available to albatrosses at the sea surface is not obvious. But the fact that the Dogger Bank in the North Sea and the Grand Banks off Newfoundland are also moraine banks favoured by both fishers and seabirds suggests a profound and consistent link between these banks and local oceanic productivity.

Surrounding the world's continents are the continental shelves, geologically part of the continental crust and covered by relatively shallow shelf seas. No deeper than a few hundred metres, penetrated by life-enhancing sunlight and often nutrient rich, the shelf seas are consistently productive. No wonder they are heavily used by seabirds and fishers alike, especially as the limited depths allow trawling. At the shelf margin, the shelf break signals the start of the continental slope which leads downward to the abyssal depths of the ocean where the seabed is generally 2–5 km below the sea surface.

The waters above the shelf break are favoured by some species. This preference is emphasized by the titles of otherwise sober scientific articles, for example 'Black Petrels patrol the ocean shelf-break' and 'Diet and feeding ecology of the Royal Albatross *Diomedea epomophora* – King of the shelf break and inner slope'.

Although once abundant on 'mainland' New Zealand, the stronghold of the Black Petrel today is Great Barrier Island, a substantial island in the outer Hauraki Gulf off Auckland. When tracked by GPS,[8] most chick-feeding petrels hastened north-east for 100–400 km, only to slow down when they reached the shelf break in waters 600–1,000 m deep. Indeed one bird headed out to the shelf-break (or slope) whereupon it swept back and forth in 30–70 km legs more or less parallel to the break. (See Map 10.) It is very difficult to resist the temptation to interpret this movement pattern as patrolling the prime feeding zone in an attempt to detect signs of food. Study leader, Robin Freeman, mused that the upwellings associated with the break may have generated high productivity, signalled by high levels of dimethyl sulphide, and thus attracted the Black Petrels.

Albatrosses breed on some of the world's remotest islands – but there are exceptions. One of the most accessible colonies is that on Taiaroa Head. Lying within the city limits of Dunedin, on the east coast of New Zealand's South Island, the Northern Royal Albatross colony is accessible by car as a Sunday afternoon outing. Possibly the thrill of the encounter is diminished by the ease of access, just as a helicopter landing on Mount Everest (if feasible) would lack the exultation of a mountaineering ascent.

For obvious disturbance reasons, most visitors are not permitted to do what the late Mike Imber did, collect pellets regurgitated by the albatrosses.[9] These pellets are chock-full of deep amber squid beaks, too horny and tough to be digested. The beaks vary from species to species, enabling an expert like Imber to identify the squid species eaten by the albatrosses. He found that the squid taken by Royal Albatrosses were species that lived above the continental shelf and shelf break. Oceanic squid species were much less prominent than in pellets coughed up by Wandering Albatrosses. Hence Royal Albatrosses became 'Kings of the Shelf Break', and exactly this predilection for scouring shelf break seas was confirmed when, a few years later, the Taiaroa albatrosses were tracked out to sea with GPS devices.[10]

Emerging from this narrative is a theme of seabirds gathering to eat wherever sea conditions are favourable for plankton growth. This is certainly true of the world's most spectacular aggregations of feeding seabirds. They may occur over the continental shelf. For example the Bering Sea in the northern summer provides feeding grounds for millions of breeding auks and further millions of Sooty and Short-tailed Shearwaters that find it worthwhile to traverse the Equator to reach the area and avoid the austral winter. Also supremely productive are the cold currents flowing from the poles roughly parallel to the continental margins. As they flow towards the Equator, so they peel away from the continent thanks to the Coriolis effect. This draws cold nutrient-rich water up from the deep, allowing plankton to flourish. That is why the Humboldt Current off Peru, the Benguela Current off South Africa and the California Current off the western United States are key fishing grounds for birds and fishers alike, a coincidence of exploitation that creates conservation problems as we shall see in Chapter 10.

High marine productivity is not exclusively associated with proximity to land. Also extremely important in generating favourable conditions are ocean fronts, where different water bodies converge or diverge. As when two Sumo wrestlers clash, so when two distinct water bodies meet, there is shuddering turbulence. The level of the sea's surface is slightly higher than 'normal' at these confluences, an anomaly that can be detected by satellite. If small planktonic creatures are forced to the surface, they or the animals that eat them may become seabird prey.

Alternatively the front may be divergent; two water bodies separate, with the result that the sea level anomaly is negative and cold waters are drawn upward from depth to 'plug the gap'. The outcome is continuing renewal of nutrients in the sunlit surface waters. Thanks to that renewal, planktonic growth flourishes. Small crustacea and their predators are nourished, and so the very animals ultimately needed to sustain seabirds thrive.

Frontal systems occur over the continental shelves and in the deep ocean. A small number of examples will illustrate the general principle of their importance to seabirds, but I will not attempt any comprehensive listing of the numerous frontal systems that help sustain seabirds across the world's oceans.

In the relatively shallow waters of the Irish Sea, the Irish Sea Front forms in late summer south-west of the Isle of Man, at the interface between an area of water stratified by weak tides and an area mixed by strong tides. Productivity remains high through the summer along the Front, and tracking has revealed how attractive is the Front to chick-rearing Manx Shearwaters. From far and wide they come. In fact, although they also make local trips close to their home colony, birds from all four British and Irish colonies studied by the OxNav group visit the Front.[11] These colonies are on Rum in the Hebrides roughly 375 km from the Front, Copeland off Northern Ireland (120 km), Skomer off south-west Wales (215 km) and Lundy in the Bristol Channel (280 km).

The major fronts of the world's oceans have an even greater impact on seabird feeding. None are more influential than the circum-global fronts that divide the Southern Ocean into roughly parallel bands of water of different temperature, just as a knitted sweater might be divided into horizontal bands of different colour. Travelling southward, the voyager will successively encounter the subtropical, the sub-Antarctic, and the Antarctic polar fronts. Each represents a major temperature discontinuity. For example, at the polar front the sea surface temperature drops from about 6°C to 2°C within a mere 40km. It is where relatively fresh cool water flowing north from Antarctica sinks below the warmer saline water of the sub-Antarctic. This sinking water generates turbulence and hence prime seabird feeding, sometimes signalled by blizzards of birds. At the maximum, densities of prions, mainly Antarctic Prions, can reach hundreds of birds per square kilometre at the Front where their crustacean prey, copepods, are concentrated. It is a zone so enticing, so nourishing, that it is worthwhile for Sooty Shearwaters

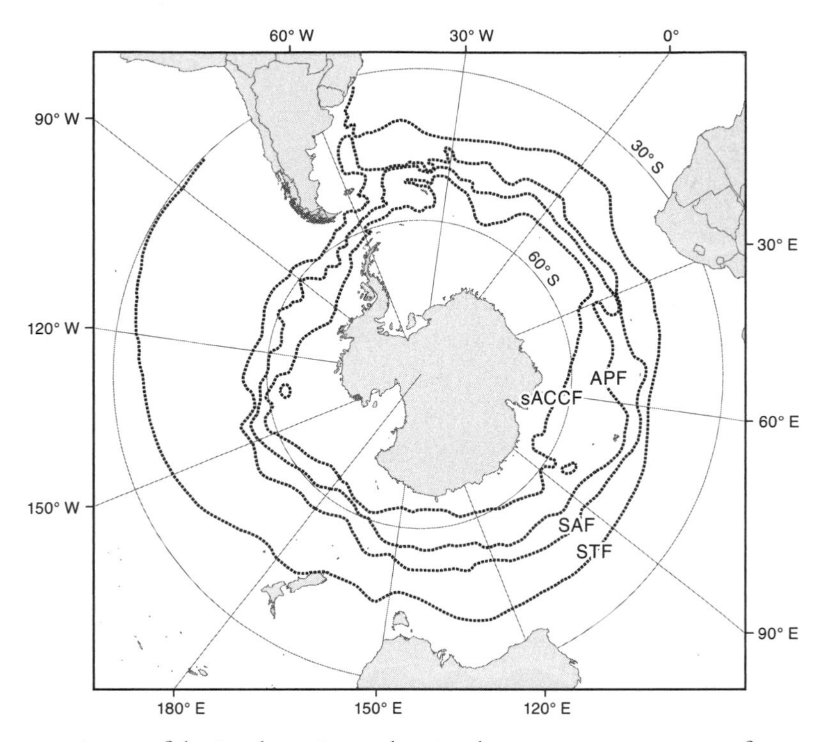


A map of the Southern Ocean showing the approximate positions of the the Subtropical (STF), the Subantarctic (SAF), and the Antarctic Polar Fronts (APF), and the Southern Antarctic Circumpolar Current Front (SACCF) encircling the continent of Antarctica.

feeding chicks on New Zealand's offshore islands to fly there, a round-trip trek of 4,000 km.[12]

Earlier I described how, during winter, the King Penguins of the Crozets head past the Polar Front, reaching as far south as the ice edge. In summer the ice edge retreats and is further from their colony, perhaps too far. Instead satellite-tracked birds that are incubating or brooding small chicks preferentially exploit the Polar Front about 400 km south of the colony.[13] The penguins swim south to the Front at speeds of 7–8 km/h. On reaching it, they slow down to about 5 km/h; they enter searching mode. This is particularly obvious where the sea surface temperature gradient is at its steepest. Then, job done, they speed up once more to 7–8 km/h for the journey home.[14]

Talk of Fronts in the Southern Ocean perhaps provokes thoughts of an ocean barrier the size of the Great Wall of China, thousands of kilometres long. That has some truth, but the Fronts are sometimes interrupted where eddies, a few tens or hundreds of kilometres across, spin off. Such mesoscale eddies can also occur where the flow of ocean currents is diverted by underwater topography. Where the eastward-flowing Antarctic Circumpolar Current* squeezes through Drake Passage, between Cape Horn and Antarctica, numerous eddies spin off to the east of Cape Horn, contributing to high regional marine productivity and hence the fabulous abundance of seabirds in South Georgia and the Falklands. However, such eddies are by no means confined to cool waters.

Off the Great Barrier Reef, the East Australian Current flows south. At the latitude of the Tropic of Capricorn, the continental shelf has an inward pocket. Into this pocket, this geological bursa called the Capricorn Wedge, spins a clockwise eddy, appropriately named the Capricorn Eddy. The spinning eddy lifts nutrient-rich water to the surface and that ultimately is why the eddy is so important as a feeding area to the 100,000 Wedge-tailed Shearwaters that make their home on Heron Island positioned on the fringe of the Wedge.

Underwater topography may not only spin off eddies at modest depths. It can also divert deep ocean currents towards the surface. Then the water reaching the sunlit surface will be cool but rich in nutrients; ideal for plankton to multiply and ultimately for bird food to flourish. Anyone who has been lucky enough to visit the Galápagos Islands will know that the two flightless seabirds, Flightless Cormorants and Galápagos Penguins, that necessarily rely on food close at hand are restricted to the archipelago's western islands. This is where the Cromwell Current, flowing from the west, hits the submarine mountain topped by the visible islands of the Galápagos. Diverted upwards, the cool current fosters an abundance of marine life. A similar effect explains feeding areas

* Also known as the West Wind Drift. Never absolutely blocked by land, the Current girdles Antarctica in a continuous flow of about 100 million cubic metres per second, an unimaginable quantity. To say it is about 500 times greater than the Amazon flow gives the imagination scant help.

favoured by seabirds in many parts of the world. For example, countless thousands of Macaroni Penguins and Antarctic Fur Seals head to a feeding area around 100 km north-west of South Georgia. Hereabouts the Antarctic Circumpolar Current is nudged towards the surface; its upwelling underpins the immense concentrations of krill that are then consumed by hungry penguins and seals.

Without question seabirds concentrate where marine productivity is highest. Where it is low, birds are few and far between. I have sailed (reluctantly) across the blue and unproductive* waters of the central South Pacific and seen one bird, a gadfly petrel, between the sun's rising and setting. Between those extremes of highest and lowest productivity, there are of course intermediate areas by no means bereft of birds. Numerous modern studies have explored where, within those areas, birds concentrate their activities, making use of the fact that satellite data are readily available for sea surface temperature and for the concentration of chlorophyll.

For example, using geolocators, a Portuguese team tracked the peregrinations of 17 Desertas Petrels during the summer breeding season when they were nesting on Bugio, lying off Madeira.[15] The birds were most active in the central North Atlantic around 40°N, about ten degrees of latitude north of the colony. Here chlorophyll concentrations were higher than near Madeira but not so high as those found another ten degrees further north. That raises the question of why the petrels, for whom 1,000 km is barely an obstacle, did not head farther north. It is a question to be re-visited later in the chapter.

The study extended into winter and the Desertas Petrels were nothing if not scattered by their pursuit of favoured areas, areas first described in Chapter 7. They were found beside the Gulf Stream off the United States, around the Cape Verdes south of Madeira, in two southern hemisphere sea areas respectively off north-east and south-east Brazil, and in the central South Atlantic. In seeking suitable seas, the birds selected areas on the west and east of the Atlantic, and north and south

* The blue clarity of the water is a direct consequence of the paucity of plankton - and life - in the surface layers.

of the Equator! All these favoured areas were characterised by relatively modest sea depths, low water temperatures (linked to the flow of ocean currents), and brisk wind speeds, aiding low cost flight.

While the petrels are apparently choosing areas of prime habitat on a huge, oceanic scale, there is evidence from other species that the choice can be made on a smaller spatial scale and on a much smaller timescale; if not on a day-to-day basis, then on a week-by-week basis. Red-footed Boobies breeding on the island of Europa in the Mozambique Channel between Madagascar and the continent of Africa like to spend the night ashore. They then make daytime excursions to catch flying fish within about 100 km of the colony. During a two-month study conducted by Henri Weimerskirch,[16] the first month was characterised by high oceanic productivity, as assessed by satellite imagery measuring chlorophyll concentrations, to north and south of Europa. Those were the directions in which the boobies headed when going to work in the morning. In the second month, oceanographic conditions changed and the accessible regions of higher productivity were west of Europa. This now become the boobies' favoured departure direction. When the sea is rather barren, as it is around Europa, it clearly benefits boobies to visit patches of temporary abundance. Perhaps they do this by following departing birds that have been successful the previous day.

While the greatest concentrations of seabirds routinely occur where ocean productivity is highest, it is clear that seabirds have the ability to detect and then exploit temporary bonanzas that are available one day - or even one hour - and gone the next in regions of lower productivity. To make reliance on temporary feasts a viable way of life, the bird must have the low flight costs that allow economical travel. The species with higher travel costs are more restricted to wherever food is assured.

* * *

When seabirds depart the colony with a feeding area in mind, the trip could well comprise a directed commuting phase, followed by a feeding phase, followed by the homeward journey. Just this pattern was evident among the Northern Gannets heading out from the Bass Rock, consistently heading either north-east or south-east (see page 133). But in comparison to the top travellers, gannets do not cover immense distances, a relatively meagre few hundred kilometres from base. Do species embarking on longer journeys obviously commute before feeding?

The answer is yes . . . sometimes. It is true of the King Penguins visiting the Antarctic Polar Front. It is also true of Indian Yellow-nosed Albatrosses, satellite-tracked during incubation. They depart volcanic Amsterdam Island, in the Indian Ocean, and head west for about 1,800 km.[17] That outward trip takes about 30 percent of the trip. They then spend 40 percent of the trip foraging in the productive waters south-east of Madagascar. Thus the area where their track is most tortuous, where area-restricted search is evident, is close to where they are furthest from mate and egg. Then, feeding accomplished, it is time to return to Amsterdam: the homeward commute takes the residual 30 percent of the trip.

The answer is no . . . sometimes. Lost in admiration for the mammoth journeys made by Henderson Island's Murphy's Petrels during 20-day incubation trips (see page 95), I cajoled my RSPB friend Steffen Oppel to put GPS devices on 10 birds in 2015. As expected, some of the trips took the birds on immense looping journeys towards the South American mainland. The birds also carried wet/dry sensors on their legs. This gave analyst Tommy Clay an opportunity to assess whether splash-downs from flight, presumably essential for food capture, were concentrated at the outer limits of the journey. They were not. They were distributed fairly evenly over the trip and so we presumed feeding happened more or less steadily throughout the entire journey. This was necessarily only an assumption since Murphy's Petrels, weighing around 400 g, are too small to carry devices that record directly when food is swallowed.

More direct evidence of actual feeding events comes from Wandering Albatrosses tracked from the Crozets archipelago. Bearing satellite transmitters and stomach temperature sensors, the off-duty incubating birds studied by Henri Weimerskirch ranged up to 2,000 km from the archipelago.[18] Each day they ate around two kilogrammes of squid, snacking throughout the journey, and both by day and night. Although there appeared to be pulses of eating, perhaps due to squid occurring in clusters, there was no obvious division of the trip into commuting and feeding phases. Consequently they foraged in a far more varied suite of ocean habitats than did the Amsterdam Yellow-nosed Albatrosses.

* * *

If certain sea areas offer good feeding prospects and others slimmer pickings, an obvious, if overly simple, question arises: why don't all birds focus their attentions on the prime areas? There are several partial answers but, even when combined, I wonder whether they provide a complete answer.

In Chapter 7, the idea of the ideal free distribution was pondered. If there are several ways of making a living, a few individuals may advantageously adopt the best. But, if increasing numbers follow that lifestyle, its profitability may drop; there are too many birds clamorously squabbling over the same fish shoal. It becomes more profitable to seek out another feeding area that is being visited by few other birds. In this way, birds may become spread out over various feeding areas so that the birds in each area fare equally well.

Potentially contributing to spatial segregation is the age of the bird. Younger birds may be less adept at feeding than their elders and betters. Better to keep out of their way.* Leading to the same outcome, a segregation of feeding areas of immature and older birds, is the fact that, at least during the breeding season, adults have to visit the colony more or less regularly to attend the egg or chick, whilst immatures without such responsibilities are relatively footloose and fancy-free. They may wander further afield. The two explanations, relative inexperience and freedom from ties to the colony, are not exclusive: both probably contribute to segregation of younger and older birds at sea.

In Chapter 3, I described how Annette Fayet discovered that, during the breeding season, Manx Shearwaters of different ages from Skomer Island travelled to different sea areas to feed.[19] Importantly the study

* Chapter 3 covered the peregrinations of immature birds. One of the outcomes of these wanderings will be a reduction in the interactions between immature and older birds, and reduced competition for food between the two groups.

was undertaken in June/July, when immatures are visiting the colony fairly frequently and are arguably as obliged to return there as are the breeding adults. Nonetheless, compared to the adults that were visiting the productive Irish Sea Front (see above p. 152), the immatures visited less productive waters and gained less weight for each day they were away from Skomer at sea. Consequently Fayet argued that the inferior competitors, immatures, were excluded from the prime areas visited by adults. Reasonable as that explanation is, it raises all sorts of questions about the mechanisms of exclusion. It seems improbable that immatures could not find the Irish Sea Front area by following tens of thousands of adult Manx Shearwaters streaming back and forth from Skomer. The question remains hanging: How does the presence of one group of birds, the adults, in the area lead to it being (more or less) shunned by another group, the immatures? Pictures of a wedding celebration come to mind, the adults and small children joyous, the teenagers unable to make eye contact on the margins.

Although the roles of male and female seabirds during laying and chick-rearing are generally similar, this is less true during pre-laying, and modern tracking has provided clear evidence that males and females, especially among the petrels, may use different sea areas at this time (Chapter 5). Another possible driver of segregation between different groups of birds is breeding failure. As soon as a bird's breeding attempt fails, and the call of the breeding colony becomes less urgent, it might pay the failed bird to forage further from the colony. Such a strategy would have the merit of taking the bird beyond the colony's zone of local food depletion, Ashmole's halo.

Helen Wade studied Great Skuas,[20] fairly or unfairly viewed as the bully boys of the Scottish Northern Isles. Once caught with remotely-controlled nooses, the skuas were fitted with solar-powered GPS trackers at two colonies in Orkney and Shetland. This enabled Wade to compare the at-sea tracks of birds still actively breeding and those whose breeding attempts had failed. The picture was satisfyingly clear; failed breeders travelled further and ranged more widely from the colony than active breeders.

These examples of different groups of birds, within a species, using different areas in the breeding season lead to the huge question of why different species consistently and predictably use different tracts of the ocean. At the extremes, the answers are trivial. Species that feed underwater by working hard and chasing the prey at depth, for example penguins and auks, need high prey densities. Otherwise their modus operandi is untenable. Underwater prey density is low in the blue water 'deserts'* of the Pacific, and there is no possibility of encountering a penguin in the immense emptiness of the central South Pacific. At the other extreme, I once travelled aboard the passenger ferry near Rum in Scotland. The ferry closed on a mob of gulls and Manx Shearwaters, and the skipper slowed the vessel as a Minke whale swam beneath the hull. Leaning over the deck rail, I could look vertically down through the clear water into the whale's blowhole. There was clearly food aplenty to attract shearwaters and gulls and whale. Where there is abundance, species can amicably share feeding areas.

But between those extremes lies the norm, where species that are modestly or trivially different in size or structure clearly use different areas. Sometimes the differences in structure do yield a clue as to why the species choose different sea areas, a pattern reminiscent of the different species gathering above tuna shoals in Pacific waters of different productivity.

Take the case of Masked and Red-footed Boobies which often nest alongside one another on tropical islands. The smaller, lighter Red-footed nests in shrubs, the larger Masked on the ground. When studied on Palmyra Atoll in the equatorial Pacific, Red-footed Boobies travelled further afield (GPS data), perhaps because of their lower wing-loading and lesser flight costs, and took more oceanic food (stable isotope data) than the Masked Boobies.[21] Thus the structural difference between booby species was plausibly associated with the use of different feeding areas but the size difference between males and females was not reflected in any comparable difference in the feeding habits of the two sexes.

Perhaps it is frustrating or perhaps it is intriguing to think of species pairs where the two species clearly have different habits at sea – for no very obvious reason. The list of potential species pairs is long. Here are

* Called deserts because the surface waters are very low in nutrients, such as nitrates. Consequently marine productivity is low, and bird food correspondingly scarce.

some examples, beginning with kittiwakes. The familiar Black-legged Kittiwake breeds around the Arctic Ocean, and further colonies dot the coasts of the North Pacific and North Atlantic. Around the Bering Sea, there is another species, the Red-legged Kittiwake, very similar but for the difference in leg colour (for which there is no ready explanation). The two species even breed alongside each other, in the Pribilof Islands for example. But there are clear differences in where they feed and what they eat. During breeding, Red-legged Kittiwakes rely heavily on lanternfish from the Bering Sea shelf break, while Black-legged Kittiwake prey on more varied fare caught above the shelf, including a larger proportion of invertebrates. The spatial differences persist through the winter. Geolocator-tracking of Pribilof birds has revealed how the Red-legged Kittiwakes remain in the Bering Sea where, in the cold and dark of winter, they use continental shelf waters and sea-ice edges. In contrast their Black-legged cousins head south and disperse widely across the sub-Arctic North Pacific.[22]

The four southern hemisphere diving petrel species are famous for their ability to fly through wave tops on whirring wings as if the water was not there. They are also notorious for looking remarkably similar, to the extent that even identifying specimens arrayed on a museum bench is far from straightforward. Two species breed on Bird Island, off South Georgia, the Common and South Georgia Diving Petrels. When tracked by geolocators through the winter, both species spent this non-breeding period in waters either around the South Georgia archipelago (mostly Common Diving Petrel) or around 3,000 km to the east-northeast (both species).[23] There was therefore some but not great overlap in the areas visited. Stable isotope data from tiny feather samples indicated that the South Georgia Diving Petrels were catching their main prey, planktonic copepods, at a slightly greater depth than the Common Diving Petrels. Yet no petrel enthusiast could confidently assert why it is the South Georgia Diving Petrels that tend to travel further afield in winter and dive deeper.

We can even extend the thought to a three-way comparison. Petra Quillfeldt used geolocators to document the winter whereabouts of the Antarctic Prions, Thin-billed Prions and Blue Petrels breeding on the Iles Kerguelen in the Southern Ocean.[24] The three species are of roughly similar weight, around 150–180 g, and similar unexciting plumage, grey above, white below. The most obvious difference is in the bill, broad in the Antarctic Prion, thinner in the Thin-billed Prion and narrowest in the Blue Petrel. On the basis of those differences, it would be difficult to predict *a priori* where the species might spend the winter. Yet Quillfeldt's results were startlingly clearcut. In contravention of their name, the Antarctic Prions spent their time almost entirely north of the Antarctic Polar Front. (See Map 12.) The Thin-billed Prions had a distribution that straddled the Front. The Blue Petrels predominantly passed the winter south of the Front. Statistical analysis showed conclusively that Antarctic Prions wintered in cool waters, Thin-billed Prions in cold waters, and Blue Petrels in the coldest. Since the diet of the three species is broadly similar, surface-caught crustacea, the reason why the crustacea caught in waters of different temperatures should be associated with different bill sizes is a lingering conundrum. When I asked Petra Quillfeldt about this puzzle, she wondered whether more northern crustacea were smaller, more appropriate for sieving by the Antarctic Prions' broader bill while perhaps the crustacea south of the Polar Front were larger, ripe to be picked up one-by-one by the Blue Petrels' narrow bill.

While seabirds focus their feeding activities on the most productive seas, some species, especially those with lower flight costs, exploit less productive areas. These differences confirm a tenet of classical ecology, that different species adopt different ways of making a living. However the differences between species may not only be in the sea areas they use. They may also apply to how they capture food, the topic of the next chapter.

## CHAPTER 9

# How Seabirds Catch Food

Some seabirds feed onshore in very obvious ways. Gulls gather at the local tip, await the arrival of the garbage truck, and then squabble with their fellows for the privilege of being the first to tear open a black garbage sack to enjoy its unsavoury contents. Far to the south, a Giant Petrel, once it has killed a Macaroni Penguin, does not delay before ripping open the victim's body cavity and thrusting its chunky bill into the warm intestines. More appealingly, a graceful Common Tern flies parallel to the beach before flicking itself in mid-air through a 180 degree turn. It has spotted a tiny fish. It splashes into the water and emerges triumphant, the prey twitching in its blood-red bill.

But modern research has shed light on the birds' less obvious feeding techniques. If frigatebirds are clad in plumage that is barely waterproof, are they able to feed without getting wet? If a guillemot is bobbing on the ocean far from land, and certainly spending the night at sea, can the night be used for feeding as well as the day? When a penguin upends and disappears underwater, how deep might it go? And, if it dives deep, how does it cope with the physiological challenges? These are the sorts of questions to which answers are now available.

Many species patrol the seas to catch their food at or very close to the surface. However, a tropical few actually specialize on catching food in mid-air. Prominent among this group are the frigatebirds. On many tropical evenings I have watched frigatebirds hanging on angular wings, sustained aloft by a breeze billowing up the cliff. Occasionally the bird's forked tail twitches to correct the bird's position after it has been buffeted by turbulence. Then, the frigatebird spots a booby returning from the sea. Instantly its languid demeanour vanishes. Thanks to deep downstrokes the acceleration appears instantaneous and, within seconds, the booby has trouble astern. The frigatebird grabs the victim's tail while the booby, head raised uncomfortably in flight, croaks. This may be annoyance, it may be appeasement. But the cruel truth is that the frigatebird will only be appeased when the booby coughs up the day's catch. It does and the frigatebird immediately forsakes harassment to swoop downward and intercept the vomited fish before it falls into the sea. In less than a minute, the booby has lost the fruits of a day's labour. The frigatebird is the victor.

While such piracy is conspicuous because it mostly happens close to shore, attacks are more likely to fail than succeed and it seems only a small proportion of the daily food of most frigatebirds is obtained by this technique. In fact, frigatebirds secure most of their daily fare far out to sea when they drop down from high altitude several times a day. Thanks to frigatebirds thrice-burdened, with GPS trackers, heart-rate monitors, and accelerometers,[1] a graphic picture of what happens during these foraging episodes is now emerging. Occupying no more than 10 percent of the day, these episodes bring the birds to within 30 m of the surface. Heart rate, around 200 beats/minute in soaring birds, increases 2–3 fold when birds are at the surface. Wing beat frequency, barely measurable in soaring birds, goes up to about four beats per second. Given the importance of flying fish in the diet of frigatebirds, it seems certain that these frantic surface forays give the frigatebirds the opportunity to grab flying fish.

Another flying fish specialist of the tropics is the Red-footed Booby.[2] When in full foraging mode, birds are continuously landing, or diving, at a rate of 30 landings or dives per hour, including 4.5 dives per hour under the sea surface. But the dives are shallow: never more than 2.4 m underwater. However sometimes accelerometers pick up dive-like movements that are not linked to submersion. These probably record the moments when the boobies are pursuing and capturing flying fish and squid at or even above the surface.

Let us lower our focus slightly downwards into the water where a multitude of species catches their food at the surface, or by shallow dives from the surface, or plunge dives from a modest height. Think prions skimming off small crustacea from the surface of the Southern Ocean, or skimmers that fly close to the surface with their lower mandible ploughing a watery furrow. When the bill strikes prey, it snaps shut by

Several species of tropical seabird, including Red-footed Boobies, are able to catch flying fish in flight.

reflex action. Only then do the birds decide whether to swallow. Or bring into the mind's eye a horde of Northern Fulmars behind a North Sea trawler gutting the catch. As the offal is flung away, so the birds can just about submerge if the effort of submersion is needed to rescue a cod liver. And who has not wondered how a tern copes with the flickering reflections bouncing off a sparkling sea as it plunges into the water to emerge from a spray of drips with a small sand eel? However, this variety of feeding techniques has been known from well before the advent of modern technology.

Nonetheless the array of modern sensors has shed new light on some aspects of feeding at or close to the surface. In particular, with wet/dry sensors, sometimes accompanied by feeding sensors, it has been possible to learn far more about when, during the 24 hours, birds are on the water, as opposed to airborne, and potentially feeding.

Over decades several strands of evidence gave credence, if not outright support, to the possibility of night-time feeding. For example seabirds living through the long winter months around the latitude of the

Arctic Circle or yet further north experience minimal daylight, hinting they must do some feeding during darkness. Among the gadfly petrels are species that have barely ever been seen feeding, perhaps because they mostly or only do so at night. And many albatrosses and petrels are primarily squid eaters. Since many squid species come closer to the surface at night – they migrate vertically on a daily basis – they could be easier to catch at night, especially if lit by their personal array of nightlights, their bioluminescence. That argument in favour of nocturnal feeding by the birds is only relevant if the squid are caught alive. If squid are caught dead, then it matters little to the bird whether they do or do not migrate vertically and whether they are bioluminescent, provided they float when dead and do not sink into the depths. And, finally, early activity recorders deployed on some more aerial species* showed they spent more time on the water by night than by day. Because they fed on the water at night, or because they simply rested on the water at night?

It would be fair to say that the topic was besmirched with uncertainty until more direct evidence became available. Early information emerged in a paper by Henri Weimerskirch and Rory Wilson with the admirably helpful title 'When do Wandering Albatrosses forage?'.[3] The answer from stomach temperature sensors was clear; 89 percent of food (by weight) was ingested by day when the Wandering Albatrosses are mostly flying, and the remaining 11 percent at night when half the time or more is spent sitting on the water. For the breeding Wanderers it turns out that hunting tactics tend to be rather different by day and by night.

Although the difference is not absolute, the daytime tactic mostly involves flying and grabbing food whenever and wherever it is encountered. The most likely distance between such encounters is 10–100 km. And it appears this Wanderer daytime strategy is shared by Laysan Albatrosses. Heading out from Oahu (Hawaii), chick-feeding Laysan Albatrosses carried camera and GPS loggers. Twenty returning birds delivered

* The less aerial species such as auks, and of course penguins, spend virtually the entire 24 hours on the water outwith the breeding season.

Carrying a camera on its belly, a Laysan Albatross from Hawaii grabs a squid in the middle of the Pacific Ocean (© Bungo Nishizawa).

28,000 images that, once collated by Yutaka Watanuki, showed the albatrosses to be feeding on large dead floating squid.* On average the birds flew 291 km between encountering squid, which were apparently detected only at short distances. Thus, after flying straight for many kilometres, the birds altered direction just a minute or two before landing at the squid. This is a very different strategy to that of the numerous seabirds reliant on finding patches of abundance, and remaining within the patch.

During the night, the Wanderers' tactics change. The birds are more likely to be on the water, floating and possibly waiting quietly for a squid to come within range of the fierce 17 cm bill innocently coloured pale Barbie-pink.[4] This contrast in hunting style is reflected in the squid caught; larger non-bioluminescent species by day, smaller bioluminescent species by night. Then around half the squid caught weigh under 100 g.[5] However, Rory Wilson, an extraordinarily inventive pioneer of the tracking technology[6] that has made this narrative possible and now a Professor at Swansea University, wondered whether the albatrosses are actually doing something smarter than merely sitting and waiting for

* Many squid species die after reproducing, the male after mating, the female after releasing her eggs.

squid. They are. Information from attached magnetometers has revealed that, in certain regions of the ocean, the albatrosses can be spinning on the water for up to six hours of the night, in tiny circles of 1.5 metres diameter. Although they cannot light up the water with electric light, Wilson thinks that, during spinning, the birds are using their large feet to agitate bioluminescent plankton, and create a light show which could plausibly attract squid – just as a squid jigger's light attracts them.

Two features of the Wanderers' lifestyle might benefit from rapid digestion. Firstly, the birds' mobility places a premium on reducing weight by digesting food quickly, and excreting surplus material. Secondly, Wanderers following ships show very little distaste for waste discharged overboard that most readers would rightly view as distasteful. To protect against the harmful bacteria associated with such food, an acidic fast-digesting stomach would be useful, the very physiological tactic of vultures. And a small study of the Wandering Albatrosses of the Crozets discovered that their stomach acidity was greater than that of other seabirds with a pH, a measure of acidity, comparable to that of vultures at about 1.5.[7]

The propensity of the Wandering Albatross to spend more time on the water by night is widely true of other albatrosses. But it is not true of smaller tracked petrels which may spend about the same time on the water by day and by night (e.g. White-chinned Petrel) or actually more time on the water by day (e.g. Chatham Petrel when not breeding). Given that the Wandering Albatross catches most of its day's food during the daytime when flying, it is sensible to wonder whether the species that are more active at night are night-time feeding specialists.

A strong candidate is Bulwer's Petrel, a rather small all-dark species found in all the world's principal tropical oceans. Studied by Maria Dias at the Salvage Islands, situated between Madeira and the Canaries, the petrels, when at sea, spend in excess of 90 percent of the night-time hours on the wing.[8] The day/night contrast is at its most extreme when the petrels, friends of the infinite deep, reach their winter quarters. These are tropical seas of the mid-Atlantic where the water is over 4,000 m deep. There, Bulwer's Petrels are 3–4 times as likely to be flying by night as by day. Since the proportion of night-time flying during winter barely fluctuated with the phase of the moon, Dias reckoned the petrels were supremely able to catch prey on the darkest nights. Very probably the fact that the principal prey, hatchetfish and lanternfish, are bioluminescent assists the petrels' hunt.

Nonetheless a follow-up study checked how at-sea activity during incubation varied with phase of the moon. When the moon rose during the night, as happens with a waning moon, the amount of time in flight increased by about ten percentage points. The increase coincided with the moon's rise, and not time of night. The reverse occurred during a waxing moon; when the moon set in the middle of the night, flight activity dropped.[9] Barau's Petrel is another species more active at night when the moon is showing.[10] It is, however, difficult to tell whether the increased activity of the birds under moonlight is because it is easier for them to catch prey, or because the prey takes precautions and is more wary about approaching the surface when the moon is visible, obliging the birds to work harder.

* * *

It is now time to venture further underwater. Quite a lot further. Of course, the maximum depth reached by seabirds varies according species, so I shall initially aim to sketch the depths attained – or plumbed – by the more adept divers among the seabird community. These depths also raise questions about how the birds cope with underwater physiological stresses that would tax or indeed kill any human diver. Finally I shall investigate when, during the dive, the prey is caught.

Despite sometimes hitting the water at up to 100 km/h, and lacking external nostrils (in order to avert the risk of water going up their noses), gannets and boobies are not especially deep divers. For example Blue-footed Booby dives average about four metres with females reaching slightly greater depths than males. That sexual difference is repeated in Northern Gannets where the females' mean depth is 4.7 m and the males' mean 3.2 m. The maxima are 18 and 11 m respectively.[11] In both species, the underwater profile of the dive can be either V-shaped, where the bird dives to maximum depth and immediately returns to the surface or U-shaped, where the bird flaps its wings and probably chases prey at the maximum depth.

Watch a cormorant or shag arch its body as it slips underwater, and it is difficult to resist admiring a bird that appears so at ease below the surface. Watch a video from a camera-carrying Imperial Shag, and the impression is confirmed.[12] After submerging off Patagonian Argentina, the bird descends for 40 seconds, its head bobbing in front of the back-mounted camera. On reaching the bottom at 50 metres, the shag begins searching, peering to left and right across a nearly featureless seabed. After another 80 seconds it catches a fish with a single lunge. Time to ascend. The light brightens, and air bubbles start to fizz forth from the plumage. At the end of the 40 second ascent, the bird emerges into South American daylight.

Cormorants and shags routinely search for prey close to the seabed at depths of 30–40 m but Imperial Shags, probably the best divers in the group, have plunged to 145 m when breeding on the Crozets.[13] Since the cormorants propel themselves underwater by feet, not wings, it might be that the Flightless Cormorants of the Galapagos can reach the same depths as their flighted cousins. This is broadly true. The maximum recorded depth is 73 m. However over 90 percent of their dives occur in water less than 15 m deep whilst within a kilometre of the nest.[14] Scope for plunging into the depths whilst remaining close to the coast is obviously limited.

The largest group of seabirds, the petrels and albatrosses, includes species that do not dive. The loose-plumaged gadfly petrels are an example. Other species may submerge but would not win any diving competition. Think, for example, of a Light-mantled Sooty Albatross splashing down to four metres. However, among two petrel groups, there are serious divers, most obviously the diving petrels. Studies from 20 years ago, using relatively inaccurate capillary recorders (Chapter 1), hinted that these birds, weighing about the same as a large thrush, could reach 40 m. More modern devices, based on solid state electronics, have yielded less startling depths. Thus the South Georgia Diving Petrel, apparently tending to hunt somewhat deeper than the Common Diving Petrel, typically dives to about 4 m during 14 second immersions, and may go as deep as 18 m.[15]

In contrast, technological improvements have not in the least reduced the impressive depths reached by certain shearwaters that are well able

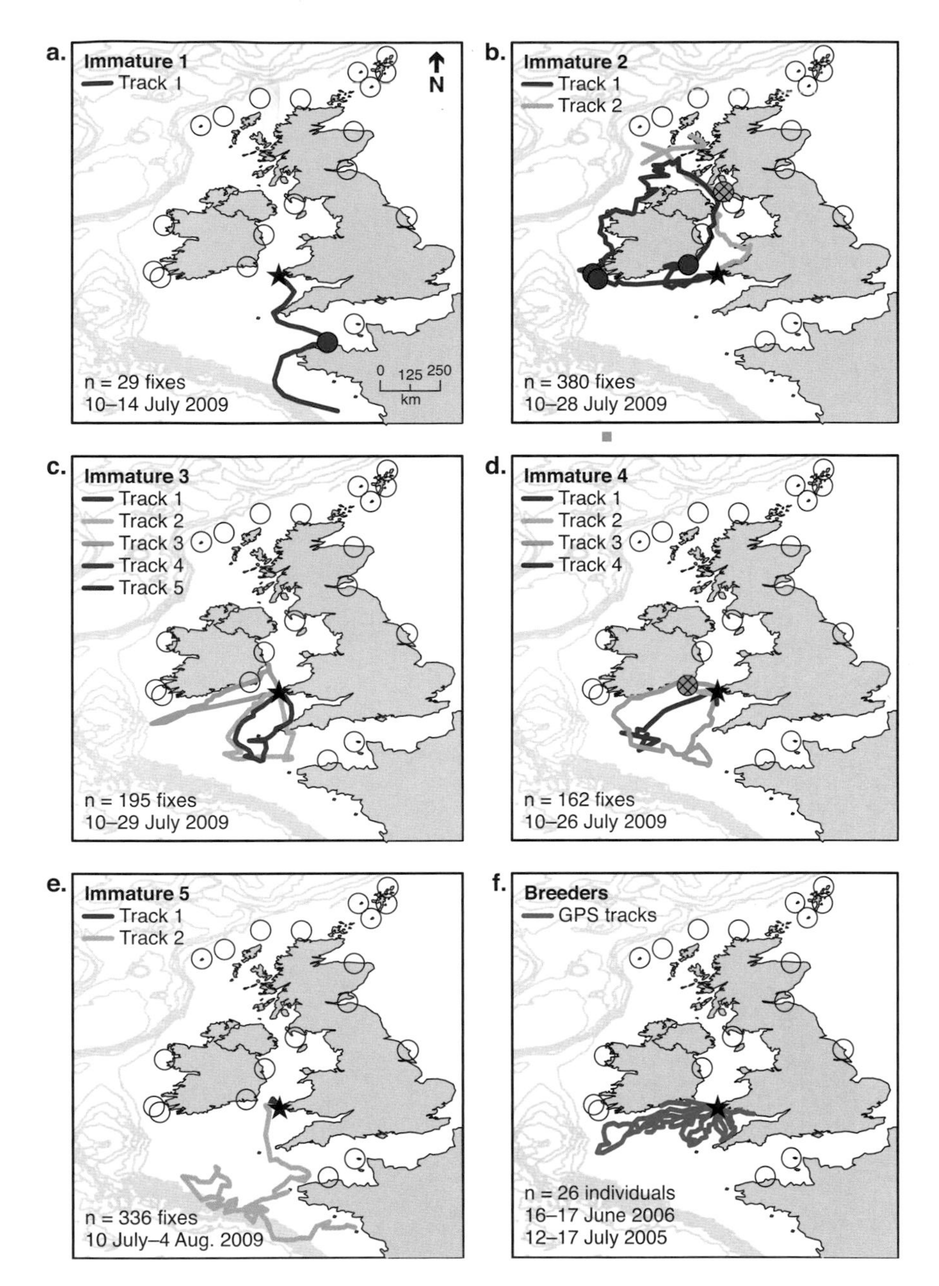


**MAP 1**. A map to show how, compared with immatures, breeding Northern Gannets tend to travel over a smaller area and do not visit other gannet colonies. Study based on GPS satellite-tracking of immature Northern Gannets caught at Grassholm, Wales. Grassholm is indicated by the black star while circles indicate all other gannet colonies in the United Kingdom, Ireland and France. Maps (a–e) show at-sea movements and presumed prospecting behaviour of five individual immature gannets caught at Grassholm 10 July and tracked until 3 August 2009. Different coloured tracks represent repeat trips after returning to Grassholm, filled coloured circles represent visits to different gannet colonies, and hatched filled circles represent birds within 10 km of another colony. Map (f) shows at-sea movements of 25 individual adult gannets (including 5 repeat tracks) breeding at Grassholm during June–July 2006. See page 46. Map reproduced, with permission of Springer, from the work cited in Note 4, Chapter 3.

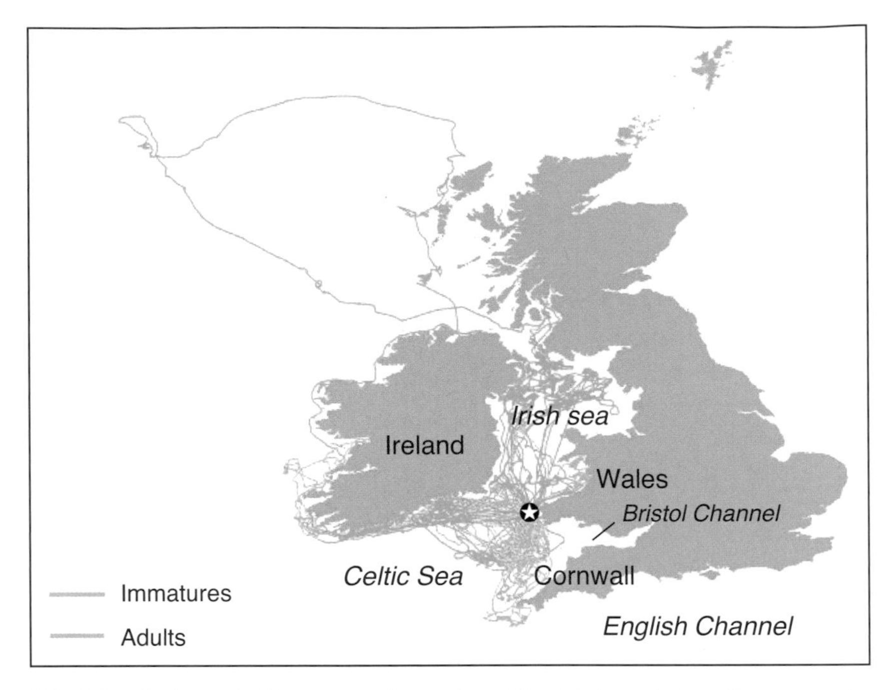


**MAP 2**. Tracks from the 20 immature (orange) and 19 adult (green) Manx Shearwaters in 2013 and 2014, respectively, visiting or breeding at the colony on Skomer Island, Wales, indicated by the star. Note how adults mostly go north, or west to waters off southern Ireland, while immatures go south. See page 48. Reproduced, with permission from Elsevier, from the work cited in Note 6, Chapter 3.

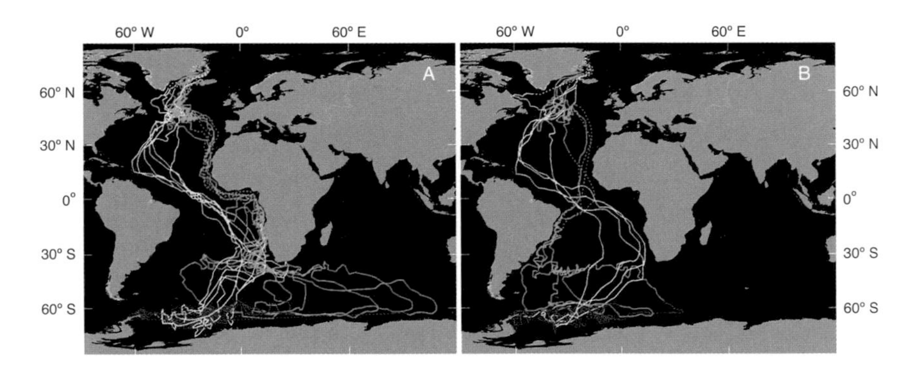


**MAP 3**. The remarkable migration tracks of 11 Arctic Terns tracked via geolocators from breeding colonies in Greenland (10 birds) and Iceland (1 bird). Green = autumn (post-breeding) migration (August-November), red = winter range (December-March), and yellow = spring (return) migration (April-May). Two southbound migration routes were adopted in the South Atlantic, either (A) following the West African coast (7 birds) or (B) following the Brazilian coast (4 birds). Dotted lines link locations either side of the equinoxes when geolocator information is unreliable. See page 55. Map reproduced, with permission from Carsten Egevang.

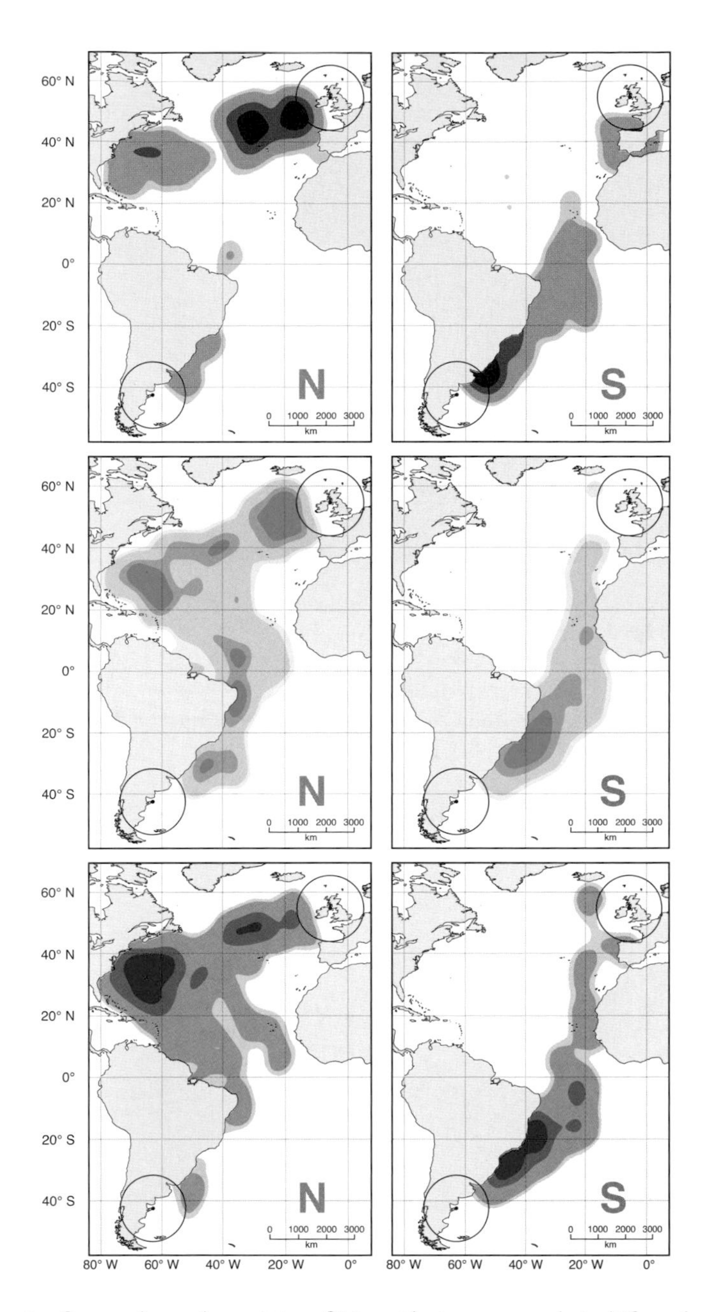


**MAP 4**. Each pair of maps shows the activity of Manx Shearwaters on their differently-routed northbound (left) and southbound (right) journeys between the Welsh colony and the wintering areas off South America. The three pairs show respectively where resting, flying and feeding are concentrated. Note, for example, how there are favoured feeding grounds off the eastern United States when birds are northbound in spring, and a resting area in the central North Atlantic. See page 70. Reproduced under the Creative Commons CC by 3.0 license, from the work cited in Note 24, Chapter 4.

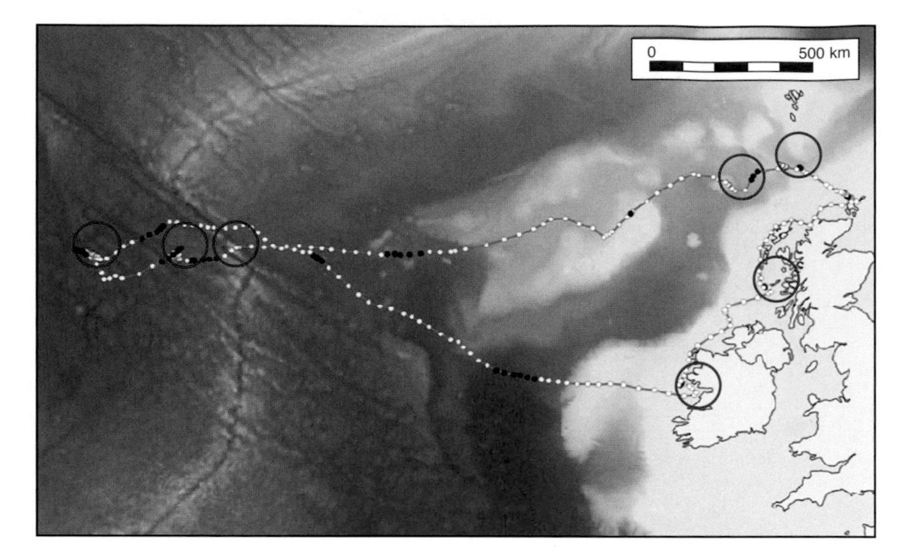


**MAP 5**. Map showing the foraging trip, lasting almost 15 days, of a male Northern Fulmar GPS-tracked from Eynhallow, Orkney, to the Mid-Atlantic Ridge and back. The map shows periods of night (dark circles) and light (white circles) on a bathymetric chart (darker colours indicate deeper water). Seven circles show portions of the journey when the flight path was more convoluted, indicative of feeding. See page 96. Re-drawn, from the work cited in Note 14, Chapter 5.

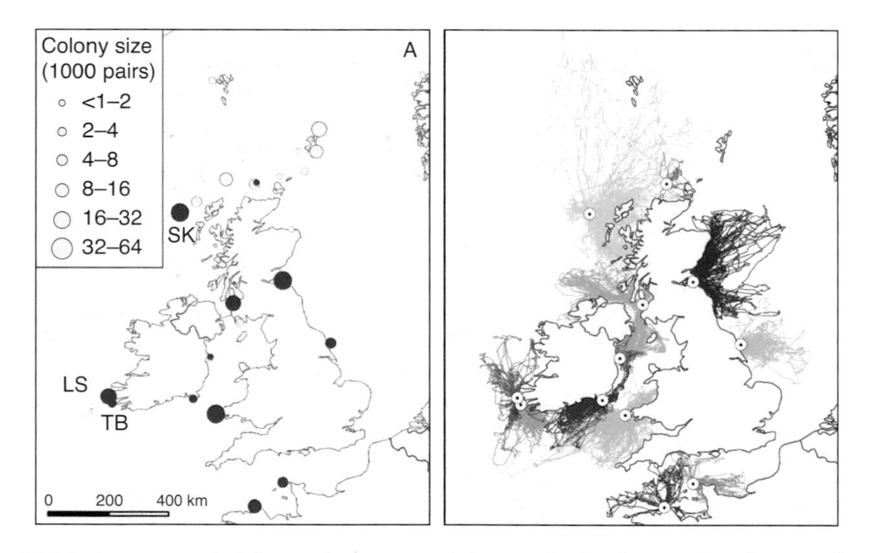


**MAP 6**. Gannets tracked from colonies around the British Isles forage in largely mutually exclusive areas, despite their potential home ranges overlapping. The left-hand map shows study colonies (red) and others (yellow); the size of dots indicates colony size. The right-hand map shows the near-exclusivity of the different colonies' foraging areas. Little Skellig (SK), Bull Rock (TB) and SK (St Kilda) are colonies mentioned in the text. See page 106. Reprinted, with permission from AAAS, from the work cited in Note 30, Chapter 5.

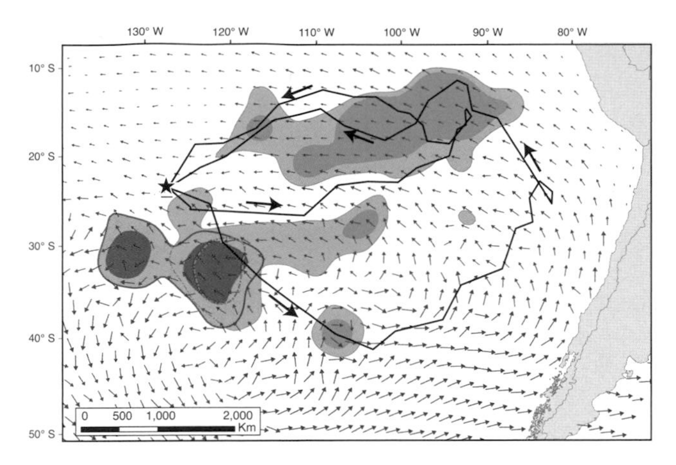


**MAP 7**. Murphy's Petrels nesting on Henderson Island (star) undertake two types of trips when off-duty during incubation. These can be to the south (orange shades, darker shading indicating more intensive use of the sea area), or looping trips east of the colony towards the Humboldt Current (green shades). Examples of two anti-clockwise looping trips are shown as black lines with black arrows indicating direction of travel. The averaged monthly wind speed and direction over the incubation period (June-July) are shown by blue arrows in the background. Arrow size relates to the wind speed, with higher speeds represented by longer arrows. These prevailing winds mean that birds on their anti-clockwise looping trips benefit from following winds on both the outward and return journeys. See page 120. Map reproduced with permission. ©Thomas Clay and Michael Brooke.

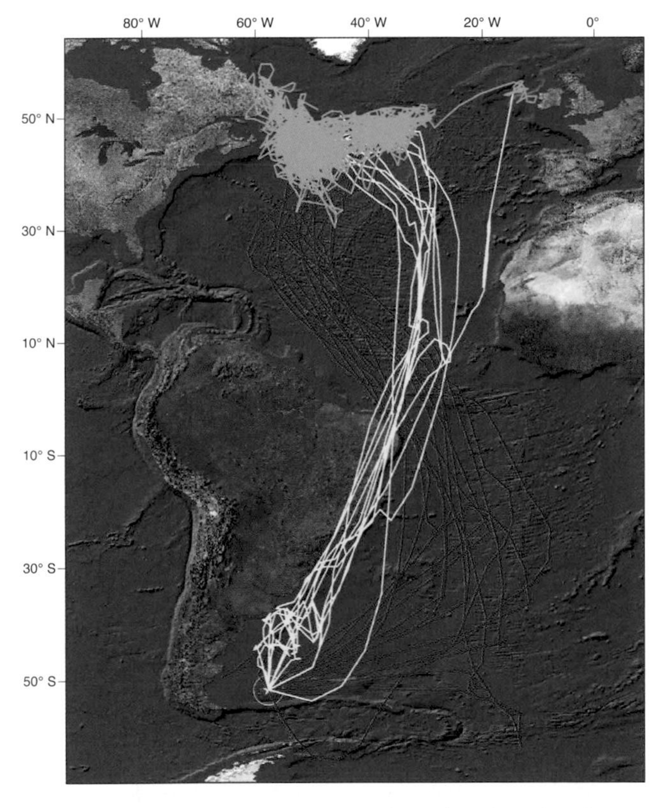


**MAP 8**. Trans-equatorial migration and non-breeding distribution of Sooty Shearwaters from Kidney Island, Falklands Islands, South Atlantic in 2008 and 2009. Red depicts the northward migration, green the main staging and non-breeding areas and yellow the southward migration. See page 124. Reproduced, with permission of Inter-Research, from the work cited in Note 17, Chapter 6.

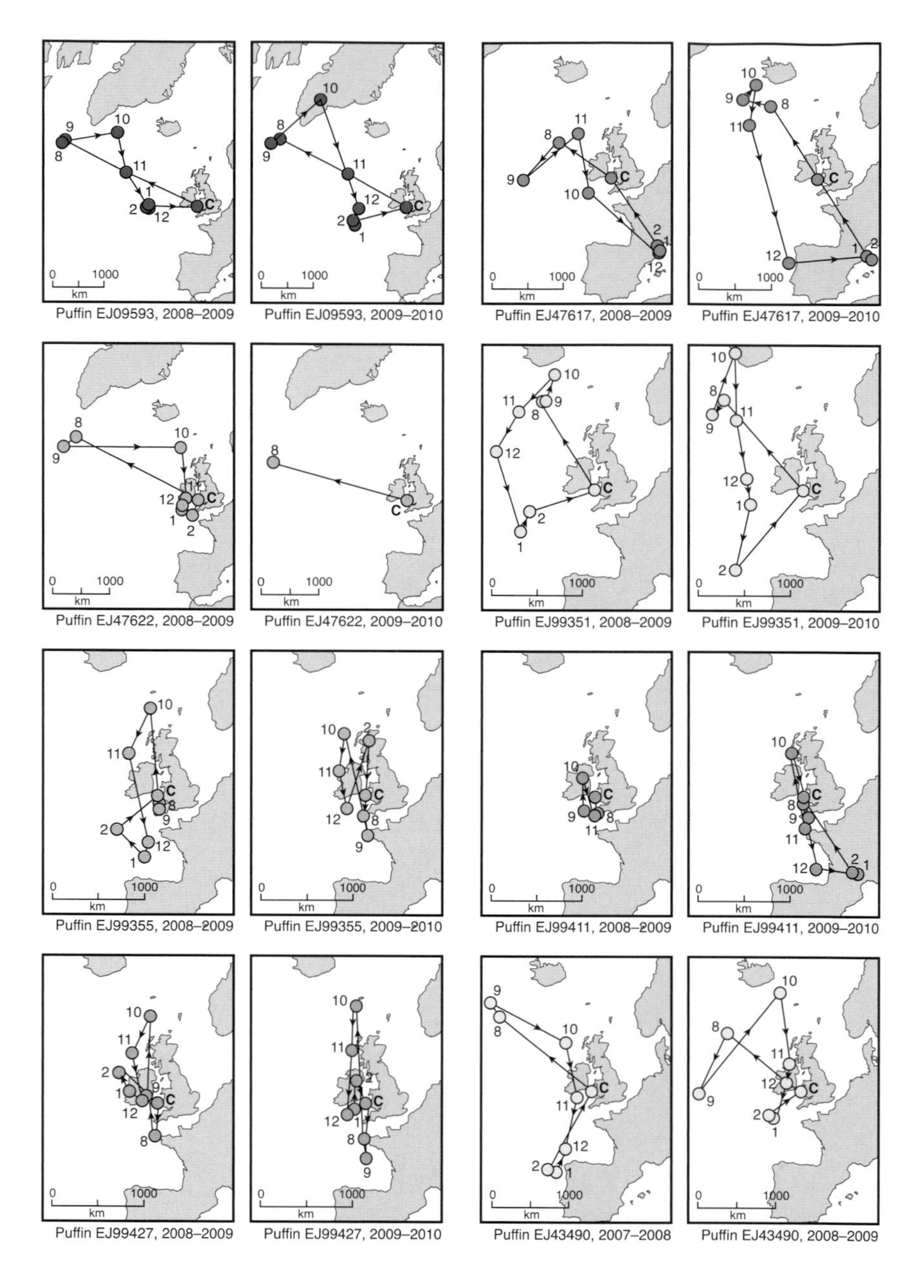


**MAP 9**. Tracks of 8 individual Atlantic Puffins during the non-breeding period in two successive years. All birds were tracked from the colony on Skomer, Wales, indicated by C. Numbers indicate the month a particular location was reached. Note how individual birds are decidedly consistent in their route choice. See page 136. Reproduced under the Creative Commons Attribution License, from the work cited in Note 9, Chapter 7.

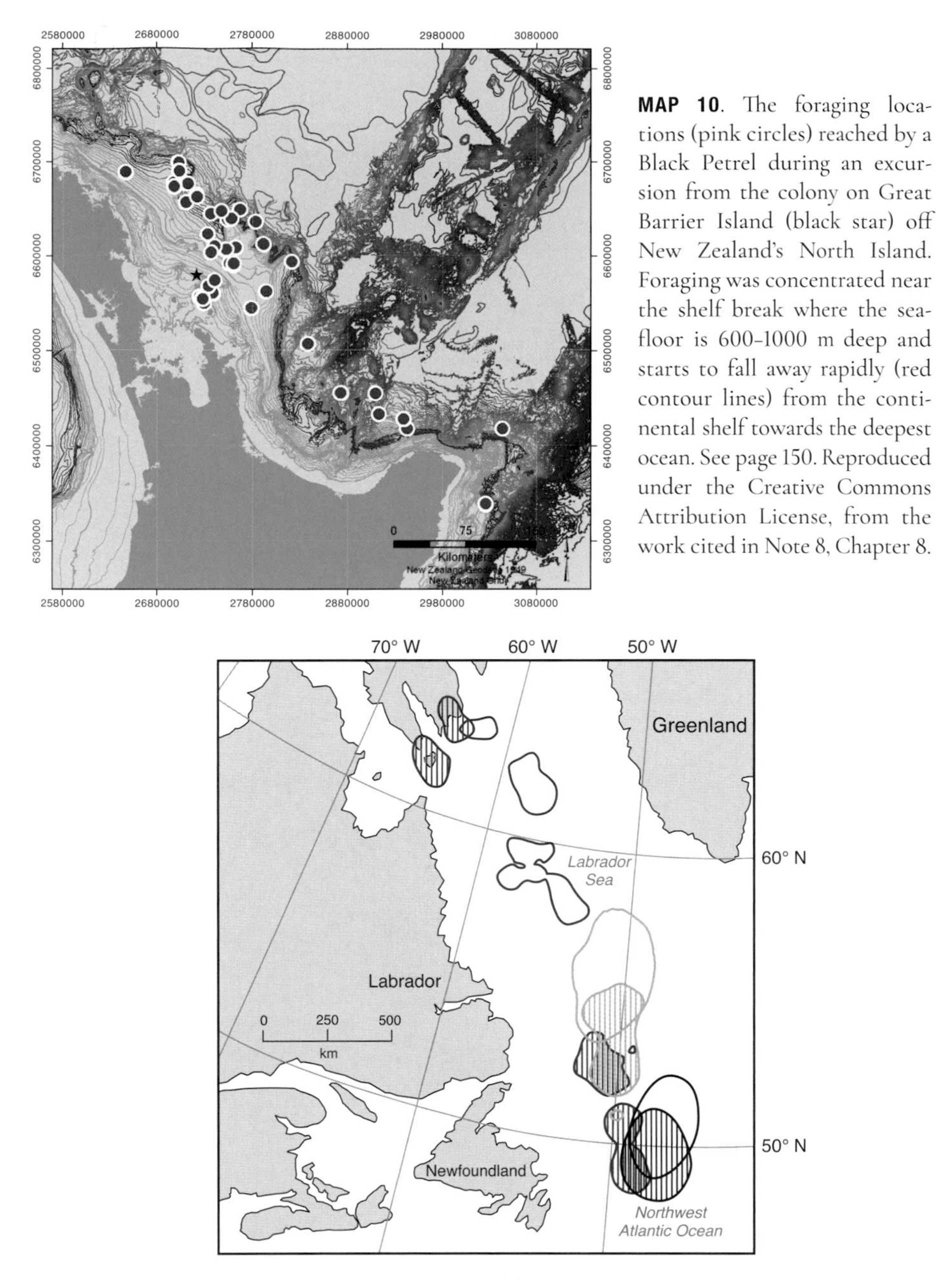


**MAP 10**. The foraging locations (pink circles) reached by a Black Petrel during an excursion from the colony on Great Barrier Island (black star) off New Zealand's North Island. Foraging was concentrated near the shelf break where the seafloor is 600–1000 m deep and starts to fall away rapidly (red contour lines) from the continental shelf towards the deepest ocean. See page 150. Reproduced under the Creative Commons Attribution License, from the work cited in Note 8, Chapter 8.

**MAP 11**. A map showing the distribution of four Brünnich's Guillemots, each represented by a different colour, across two winters, one unshaded, the other hatched. The birds bred in colonies to the north of the mapped area. There are two examples of birds that used very similar areas from one winter to the next (black and orange), and two more that used very different areas (blue and red). See page 139. Re-drawn and reproduced under the Creative Commons Attribution CC by 4.0 License, from the work cited in Note 11, Chapter 7.

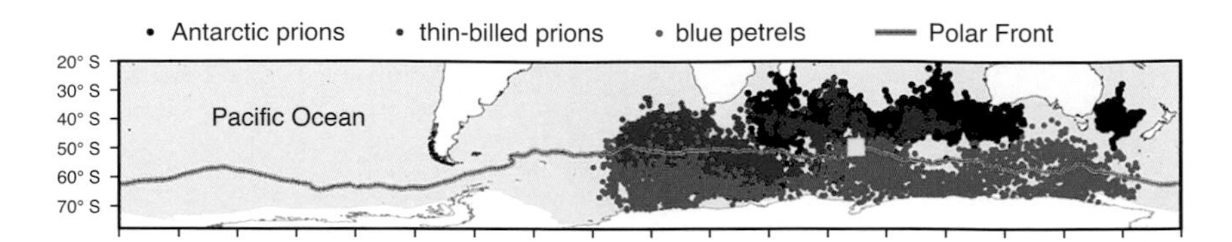


**MAP 12**. The largely non-overlapping non-breeding distributions of two prion species and the Blue Petrel from Iles Kerguelen, shown by yellow square. The black geolocator positions derive from Antarctic Prions (10 birds: 2,916 positions), red positions from Thin-billed Prions (15 birds: 3,143 positions) and blue positions from Blue Petrels (12 birds: 2,237 positions). Also shown is the position of the Antarctic Polar Front. See page 162. Reproduced, with permission of The Royal Society, from the work cited in Note 24, Chapter 8.

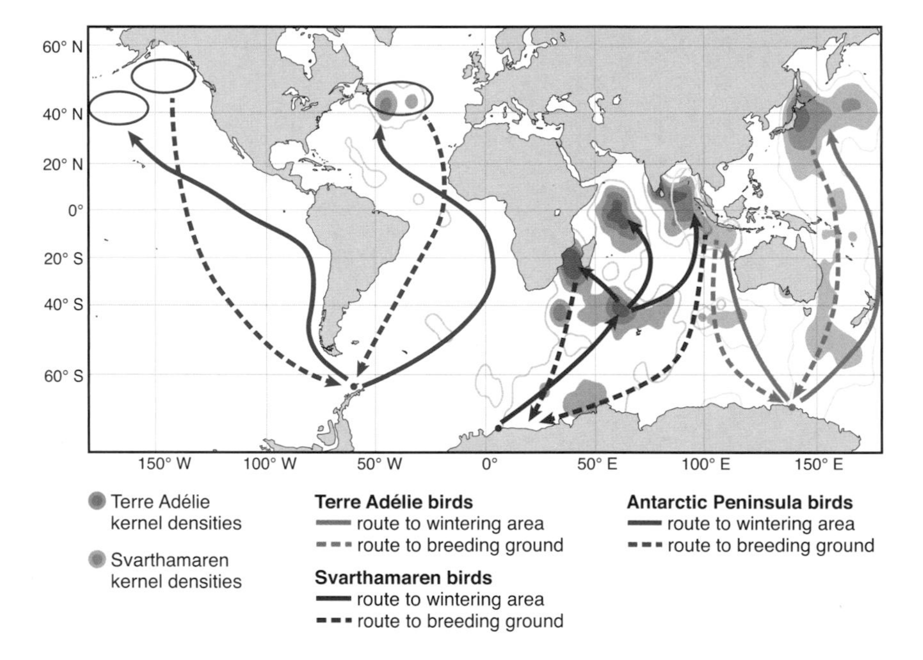


**MAP 13**. The divergent routes of Antarctic peninsula and Terre Adélie South Polar Skuas, respectively east and west of South America (blue arrows) and east and west of Australia (green arrows). Birds from the Svarthamaren colony close to 0°E are not mentioned in the main text, but they also migrate to two different oceans to escape the Antarctic winter, although most go to the Indian Ocean. See page 65. Re-drawn from the work cited in Note 15, Chapter 4.

to complete hundreds of dives per day when absent from their colonies. To pluck a couple of examples, the maximum depths reached by Manx and Sooty Shearwaters are 55 and 70 m, although the average depths are far less, respectively 10 and 16 m.[16] In general, the heavier a shearwater is,* the longer it can remain underwater and the deeper it can dive. Provided the comparison is made between similar creatures, this is true of other seabirds, and indeed more widely of marine mammals. The reason is that the larger animals have greater oxygen storage capacity while their metabolic rate per unit weight is lower – which allows the oxygen taken underwater at submergence to last that much longer.

Within that framework, the pattern shown by the auks, all of which are divers, broadly follows expectation. The Little Auk or Dovekie, weighing roughly the same as the diving petrels, reaches a maximum depth of about 25 m and remains underwater for a maximum of 1.5 minutes. The largest extant auks, Common and Brünnich's Guillemots, weighing about one kilogram, can dive to 80–90 m[17] or even 170 m,[18] and remain underwater for 2–3 minutes. We can be confident that those values would be comfortably exceeded by the extinct Great Auk!

The underwater feats of shearwaters and auks pale in comparison to those of the penguins where the relationship between diving ability and size is again apparent. The smaller penguin species may be able to remain underwater for around five minutes and reach a depth of 100 m. It is the two largest species that amaze. Early hints of their abilities came from pioneer work in the 1970s and 1980s by Gerald Kooyman of California's Scripps Institution of Oceanography. Now we know King Penguins can reach over 300 m, but they must concede pride of place to Emperor Penguins. In 2007 Barbara Wienecke and colleagues from the Australian Antarctic Division published a paper with a title 'Extreme dives by free-ranging emperor penguins'. It is a title that smacks of an extreme fishing programme on TV, and yet it is a sober report of science on a near-industrial scale. I can only quote from the paper's abstract. "We examined the incidence of extreme diving in a 3-year overwintering study of emperor penguins *Aptenodytes forsteri* in East Antarctica. We defined extreme dives as very deep (greater than 400 m) and/or very

* Sooty Shearwaters (800 g) weigh about twice as much as Manx Shearwaters (420 g).

Emperor Penguins are probably the deepest-diving of all seabirds, sometimes plumbing depths in excess of 500 m in pursuit of fish and squid.

long (longer than 12 min). Of 137364 dives recorded by 93 penguins 264 dives [0.2%] reached depths greater than 400 m and 48 lasted longer than 12 min [0.03%]. Most (65%) very long dives occurred in winter (May–August) while 83% of the very deep dives took place in spring (September–November). The two most extreme dives (564 m depth, 21.8 min duration) were separate dives and were performed by different individual penguins. Penguins diving extremely deeply may have done so as part of their foraging strategy whereas penguins diving for very long times may have been forced to do so by changes in the sea-ice conditions."[19] No wonder Wienecke's face is one characterised by raised eyebrows expressing surprise and delight at the dives of penguins, and life in general.

Once information about these impressive depths is to hand, all manner of questions surface. One is how the birds detect prey before actually catching it. In some cases, there is little doubt that vision is the primary sense. I have already mentioned the Imperial Shag swinging its head from side to side as it scours the sea bed. Among penguins catching krill, there is a coincidence between catching individual krill, as recorded by back-mounted cameras, and side-to-side head movements as recorded by accelerometers, exactly as one would anticipate if the birds were spotting and then picking off the krill one by one.[20] In early 2016, Jonathan Handley, a doctoral student at South Africa's Nelson Mandela Metropolitan University, attended the Student Conference in Conservation Science at Cambridge University. He played video clips obtained from cameras back-mounted on Gentoo Penguins going about their underwater fishing business in the Falklands. When the penguins caught rock cod and lobster krill, the audience burst into spontaneous applause. Everything indicated that the penguins used vision to find those prey underwater, as was true in a remarkable instance of underwater piracy also recorded by Handley – or more accurately by the cameras aboard his penguins. When one penguin caught a large squid* at a depth of about 30 m, a second and then a third penguin joined the fray before the captor could swallow its prize. In the melee, the squid was torn apart and shared.[21] Again it seemed overwhelmingly likely that the pirates saw the squid caught by their fellow penguin, and promptly abandoned good manners.

Using sight to detect prey underwater obviously requires light. Yet, even on a bright day, minimal light penetrates clear water below about 200 m, which is consequently the absolute lower limit of photosynthesis. Of course the waters within 200 m of the surface are darker still by night, and yet there is little doubt that some birds are feeding underwater at night.

Take the case of Common Guillemots collecting food, such as spawning Capelin, for their chicks. Diving by day to 100 m off Newfoundland they encounter light levels equivalent to moonlight, but rarely dive to depths where the light equals starlight. Underwater cameras reveal that, by day, the Capelin caught by the guillemots are mostly isolated individuals, not those in denser shoals.[22] That is compatible with visual detection.

At night the picture alters. Although the Guillemot dives are rarely deeper than 50 m, the birds are nevertheless active in the equivalent of

* The length of the squid's mantle, roughly the body, was 15 cm.

starlight, conditions they would eschew by day.[23] Whether the birds have the visual abilities to detect prey from afar in such conditions is uncertain. The alternative possibility is that the Capelin are sometimes packed in such tight shoals that random encounters are sufficiently frequent to make the guillemots' dives worthwhile. But I do admit that the idea that a guillemot might dive at night in the hope of bumping into a fish sounds bizarre. Possibly the idea becomes less bizarre when one remembers that the water temperature at 50 m off Newfoundland is close to freezing, chilly enough to render Capelin somewhat sluggish.

While any tendency by seabirds to dive less deeply when it is darker could be driven by light – or its absence – another factor might be vertical migration of the prey. If the prey are closer to the surface at night, it would be unnecessary to then dive to daytime depths. And Capelin do indeed migrate, concentrating around 20 m below the surface by night, and sinking beneath 200 m by day.

Also migrating vertically are the lanternfish prey of King Penguins which, like the Guillemots, dive less deeply by night. During the night, shallow dives under 30 m are the norm for King Penguins, whereas deep dives, typically to 100–200 m, occur only during daylight.[24] However, lanternfish are well scattered during the night, and so any random approach to capture would be unlikely to work for the King Penguins. But the lanternfish, unlike Capelin, are bioluminescent, so it may be that individual fish can be detected from afar by King Penguins.*

Diving to any sub-surface feast necessarily poses problems. An obvious problem is that, when underwater, the bird cannot breathe and must eke out those oxygen stores with which it submerges for as long as possible. Since this is not a physiology book, I shall only sparsely outline the adaptations that help the diving bird overcome the difficulties. These adaptations have been most extensively studied in Emperor Penguins. Given its size, and therefore predictable rate of oxygen consumption, an Emperor Penguin could remain underwater for about five min-

* Lanternfish are also a main prey of Southern Elephant Seals that share many a sub-Antarctic beach with King Penguins. The peak sensitivity of the vision of female Seals is at the very wavelength at which lanternfish shine. http://dx.doi.org/10.1371/journal.pone.0043565 (accessed 14 June 2017).

utes if its body processes continued to function as they do when it is breathing air. This 'limit' is comfortably exceeded by recorded dives lasting some 20 minutes. Just as a breathless athlete striving for the finishing line builds up lactic acid, so the underwater penguin builds up lactate, principally in the muscles. This is then flushed out when it returns eventually to the surface.

Another key adaptation to diving is a reduction in heart rate underwater, exactly as also occurs in diving seals and whales. Detected via attached electrocardiogram (ECG) recorders, the heart rate of a resting Emperor Penguin is around 70 beats/minute. This value roughly doubles immediately before the dive. If the dive is short, under five minutes, the underwater rate is about the same as when resting. If the dive is long, heart rate drops off dramatically, and may reach as low as three beats/minute. Just before the penguin surfaces, the rate accelerates. It can be around 200 beats/minute when the penguin surfaces and can breathe once more to replenish its oxygen stores.[25]

Remembering that even in tropical seas, the water temperature below 200 m is probably no higher than 5°C, a further physiological problem faced by seabirds underwater is potentially that of cold. Penguins and auks have tight plumage that retains air close to the skin. This assists heat retention, albeit by creating buoyancy that hinders the downward dive. The situation is different in cormorants and shags. Their plumage is notoriously wettable. Think of the classic pose of a perched cormorant hanging out its wings to dry after a spell of swimming. If the water has reached the skin, the cormorant will have lost more heat than another seabird whose skin remains dry.

In fact the paradox is more apparent than real. The Great Cormorant is one of the most widespread of all seabirds. Its breeding range extends from Australia via south-east Asia to Europe and Greenland. Whether in northern France or Greenland it requires less food per day than other seabirds of comparable weight.[26] The birds are evidently not leaking heat, and not needing to take in extra food. How they retain heat became evident when European researchers looked at the plumage more closely. All four subspecies studied, living in sub-Arctic to subtropical climes, retained an insulating air layer in their plumage, which was, however, much thinner than for other species of diving birds. Detailed examination of the plumage showed that each cormorant body feather has a loose, instantaneously wet, outer section and a highly waterproof central portion.[27]

There is another cormorant species we have already met, the Imperial Shag that is confined to cool waters of the Southern Hemisphere. It, too, has a waterproof inner plumage that traps air and hence heat. The trapped air creates less buoyancy and is therefore less troublesome on deeper, as opposed to shallower dives. That is probably the reason why Imperial Shags are deeper divers than most other cormorant species of lower latitudes.[28]

Because of air trapped in the plumage, plus that in the lungs and air-sacs, seabirds are positively buoyant when first submerging. Their natural tendency is to float back upwards. Therefore they have to work hard to reach greater depths. As the birds reach those greater depths, so the air compresses and eventually a depth is reached where they are neutrally buoyant. This depth depends on the quantity of air in the lungs and airsacs at submergence; the more air to be compressed, the greater the depth of neutral buoyancy. In fact penguins anticipating diving deeper take in more air before submerging which then allows a longer dive and increases the depth at which they are neutrally buoyant. Therefore, rather neatly, there is a close correlation between diving depth and duration, as was evident from the stand-out numbers earlier in this chapter.

Having, in Rory Wilson's evocative descriptions, pedalled downwards, the bird faces an ascent that is relatively easy free-wheeling. This is simply because the compressed air expands at shallower depths, increasing positive buoyancy and serving to thrust the bird upward at an ever-faster speed. For example, Magellanic Penguins beginning their ascent from 50 metres initially ascend at a little more than half a metre per second. When they have reached 20 metres, the ascent rate has topped a metre per second.[29]

This rapid freewheeling ascent has two consequences. The first is that the bird often emerges rapidly onto the surface, like the proverbial cork popping out of the bottle. This is of no particular consequence if the bird is in mid-ocean. It is far more significant when the penguin emerges onto ice – and familiar indeed are TV sequences of penguins apparently

As a submerged penguin approaches the surface, air trapped within its plumage and lungs expands, increasing the bird's buoyancy and rate of ascent. This impetus helps Gentoo Penguins, and other species, leap onto ice.

'exploding' out of the sea to jump a metre or more onto the security of ice beyond the jaws of a Leopard Seal. To facilitate a clear run at an isolated exit hole through ice, Emperor Penguins that have been feeding at the undersurface of ice descend to depth before returning to the exit hole.[30] Of course, sometimes the Penguins are embarrassed by a failure to achieve the necessary height and, amid whirring flippers, splash back shamefacedly into the water. In a study of nine instrumented penguins, the maximum swim speed in the second before exit correlated with the above-water height of the ice surface, indicating that the penguins anticipated the ice height. Nevertheless there were 37 failures among 386 leaps. Those failing penguins did not seek out lower holes after failing to exit through the higher holes. Rather, swim speed, typically around 3 metres a second, was increased for subsequent attempts. And the researchers, watching their subjects from a sub-ice observation chamber, could see that some exits were achieved with little or no flipper beating; the positive buoyancy generated adequate speed.[31]

If diving seabirds tend to be freewheeling on their ascents, this may give them the extra agility that aids grabbing prey. In contrast, the hard work of pedalling downward might hint that the descent is not the ideal time for hunting and catching. In fact two contrasting approaches are now yielding information on when, in the course of dives, birds actually catch and ingest food. The first are beak-mounted sensors that detect how widely apart the upper and lower mandibles are. Opening the bill is obviously essential for catching prey, and the bigger the prey, the more widely the bill must open. The second is the attachment of accelerometers which register the bird's movement in three dimensions. Prey capture is likely to be linked to sudden movements, contrasting with the relatively steady movements of, say, the descent phase of a dive. If information from the two channels can be combined, so much the more informative.

Rory Wilson achieved this combination with Magellanic Penguins. Around 90 percent of prey was captured during ascents. These were not necessarily the principal ascents back to the surface. More often they were small upsurges or undulations when the penguin was at depth in the middle of a dive with an overall U-shaped profile. Moving in for the kill from below the prey may not only give the penguin an advantage in buoyant mobility but it may also be easier for the bird to spot its potential catch when that prey is silhouetted against the lighter background overhead.

The level of detail that can be gleaned from accelerometer data is remarkable. For example, Marianna Chimienti of Aberdeen University attached the devices to Razorbills and Common Guillemots breeding at Scottish colonies.[32] Razorbill dives were consistently V-shaped in profile, and relatively shallow, to no more than about 10 m. Prey was predominantly caught during ascents. This underwater behaviour was clearly different from that of the Guillemots whose dives were of two types. They undertook shallower dives, to 10–30 m, where prey was caught either at the greatest depth or during the ascent. Alternatively the Guillemots went deeper, to around 50 m. The profiles of these dives were clearly U-shaped. During the bottom phase of roughly constant depth, the birds were either searching for or catching prey.

It almost defies belief that the seabirds we see in two dimensions, essentially at or very close to the sea surface, are actually catching food and making a living in three dimensions. From the Great Frigatebird 1,000 m high in the tropical sky to the Emperor Penguin at a frigid depth of 500 m, different species are actually occupying three dimensions. Their capacity to thrive in the marine environment is, once more, more remarkable than we ever anticipated.

## CHAPTER 10

# The Clash

Seabird Interactions with People – Past, Present and Future

Despite their truly remarkable ability to forge a living at sea, whether in sight of land or at an immense distance over the horizon, seabirds are not thriving globally. They are arguably the most threatened group of birds, because of a combination of hazards met ashore during breeding and perils encountered at sea.

Ashore, seabirds have been harvested by people since time immemorial. Colonies have offered a bonanza of flesh, feathers, and eggs that has proven irresistible to communities living near the coast. Think of the iconic photographs, from a century ago, of the bearded menfolk of St Kilda in solemn contemplation of a pile of fulmar corpses.[1] While such harvesting continues, it is, from a worldwide perspective, no longer a primary cause of seabird decline. More significant is the continuing devastation wrought by alien species introduced to seabird nesting islands. The impacts may be direct and brutal. Introduced cats and rats eat seabirds, their chicks, their eggs. Once thriving colonies have been massively reduced or totally extirpated. The main islands of New Zealand, formerly mammal-free (except for bats), once held huge seabird colonies. These have gone forever thanks to the impacts of Maori and then European settlers, their attendant mammals, and associated wider habitat changes.

Focussed examples are more vivid still. In order to be stripped of their furry skins and provide high value coats, blue foxes were introduced to more than 400 islands off Alaska from the early 18th century onwards. The foxes devastated seabird colonies, notably auk colonies. As they colonised Polynesia, so the Polynesians introduced Pacific Rats to numerous islands, sometimes accidentally, sometimes deliberately since they had no cultural aversion to a lunch of grilled rat. I have watched the consequences on Henderson Island where the Pacific Rats, the Polynesian legacy, routinely snatch a one-day-old fluffy petrel chick from beside its parent. To its shame, the parent makes scant effort to defend its young. Once dragged a metre away, the still-living chick continues to cheep weakly as it is chewed open and eviscerated by the rat. And, very startlingly, mice on Marion Island in the Southern Ocean, accidentally introduced to the island by sealers, have recently been reported to 'scalp' albatross chicks.[2] The mice, which have probably developed this 'skill' only recently, clamber onto the neck or head of the young albatross where they are safely beyond the reach of the chick's snapping bill. There they begin to strip the albatross's flesh, resulting in a bloody scalp and sometimes causing the chick's death.

Other introduced species can affect seabirds ashore via indirect pathways. In the Juan Fernández Islands off Chile,* rabbits and goats contribute to erosion, thereby reducing opportunities for petrels to burrow.

* Goats were essential for the lifestyle of Robinson Crusoe, whose story Daniel Defoe was inspired to write after hearing about the real-life adventures of Alexander Selkirk, marooned at his own request on the Juan Fernández Islands from 1704 to 1709.

On the Australian sub-Antarctic island of Macquarie, rabbits caused immense damage to the island's vegetation and seabird habitat. Over a span of almost 50 years, their eradication required the introduction of fleas followed by the myxoma virus and then the application of 300 tonnes of poison bait. Undoing the damage caused on islands by alien species is by no means easy.

Nevertheless, projects to eradicate these alien mammal species are now numerous; over 1,000 have been undertaken. Most are successful and, provided sensible quarantine is observed, offer the prospect of a conservation gain in perpetuity. Their execution can range from the apocalyptic shooting of goats from a helicopter, the technique used to rid northern Isabela in the Galápagos of over 100,000 goats, to the more high-tech. Rat eradication on larger islands now regularly uses helicopters guided by GPS to broadcast swathes of poison bait pellets. The precision guidance ensures that, almost literally, no square metre of ground fails to receive a pellet from the sky.

Such island restoration ventures are a boon to seabirds but are generally achieved without the benefit of the extra knowledge of the seabirds themselves that has come from modern gadgetry. However onshore conservation does make use of this gadgetry in unanticipated ways.

One onshore conservation problem is light pollution. In particular the young of seabird species fledging at night are attracted to lights ashore,* especially during new moon periods. Instead of flying or swimming out to sea, the youngsters find themselves in unexpected and unsuitable places on land. In the Outer Hebrides, young Atlantic Puffins on St Kilda wander into the island settlement on Hirta. A little to the east, towards the mainland of Scotland, Manx Shearwaters fledging in autumn from the island of Rum should be bound for the South Atlantic. Any bird rescued from the streets of the nearby fishing port of Mallaig is definitely confused. Fledging Hornby's Storm Petrels appear in thoroughfares of the northern Chilean city of Antofagasta in July every year, to be rescued by a local conservation group. This seasonal fall-out

* Just why nocturnally-active seabirds should, like moths, be attracted to light is not entirely clear – but the fact is they are.

is wholly predictable and so tells us when this mysterious species breeds (see Chapter 1).

Light pollution is an especial hazard for the Cory's Shearwaters of Tenerife. The largest of the Canary Islands, Tenerife, with a human population of about 900,000, is a destination for some 4.5 million tourists annually, and provides nesting space for at least 2,500 shearwater pairs. Possibly as many as half the fledglings reared each year are grounded by light attraction as they leave their burrows and head to sea by night. This first journey tends to be made in the first three hours after sunset, the very time when residents and holidaymakers are most likely to be wining and dining under artificial light. Twelve youngsters GPS-tracked from colony to touch-down all came to earth within 10 km of the colony but the tracks revealed that some had overflown the sea before returning to land.[3] Thus the young shearwater is not secure from the lure of light once it is over the sea. To reduce the problem, light levels need to be brought down, not only in the vicinity of colonies, but along the Tenerife shoreline.

Thanks to rescue campaigns, usually conducted by volunteers, over 130,000 birds worldwide are given a second opportunity to reach the sea after tumbling out of the sky. But I fear the problem will persist for many years. The several Save Our Shearwater campaigns, a name* that has gained traction in various parts of the world for very obvious reasons, will need to remain active.

Another onshore hazard is posed by power lines. The scale of the collision problem is immense. Every year, there may be as many as 25 million bird/line collisions in Canada. And the problem is exacerbated for seabirds because many of the species nesting a few kilometres from the sea journey to and from the colony in darkness when, of course, the collision risk escalates. For millennia, Newell's Shearwaters and Hawaiian Petrels have crossed the dissected volcanic coastline of Kauai at dusk and headed inland, gaining height and crossing green steep-sided valleys on the last leg of their trek to the colony. Nowadays power lines may

* The equivalent, Sauvons les Pétrels, on the Francophone island of La Réunion lacks a certain *je ne sais quoi*.

stretch across those valleys from one misty ridge to another. Alas, the peace of the tropical night is sometimes broken when birds slam into lines.[4]

To mitigate this problem, good information on the number and spatial distribution of collisions is needed. Searching for downed birds in the thick vegetation bristling out of near-vertical terrain under the lines is beyond impractical. But pioneering work has shown that modern song meters may help identify collisions. Set beside a power pylon, the meters record hundreds of hours of sound onto a memory disc. Mostly the recorded sound is the rustle of the Hawaiian wind, but occasionally it is the sound of a bird hitting the wire. Since nobody can listen to the playback of sighing wind for hundreds of hours, and remain awake, the trick is to develop computer software that will pick out the sound of a bird collision from the prevailing background noise. Programmers are currently addressing this very problem. In due course, such work may be a precursor to pinpointing factors which make particular lines more or less dangerous and, eventually, to developing solutions.

When in transit between their colonies and their feeding areas, more or less far offshore, seabirds frequently gather en masse. Great Shearwaters assemble in their hundreds of thousands off Nightingale Island adjacent to Tristan da Cunha. Below the famous Bempton Cliffs of Yorkshire, England, thousands of Common Guillemots and Razorbills gather to rest, wash, and preen. It defies belief that jet skiers should choose to open the throttle and rampage through the flocks - but they do.[5] Since these areas close to colonies fall within national territorial waters, the passage of the necessary conservation legislation is not hindered by international issues.

* * *

Most seabird species spend the great majority of their time at sea, often far offshore and out of sight of land. Out of sight does not mean out of danger. Chemical pollutants reached Antarctica from the industrial world decades ago. Oil slicks and the associated images of birds' struggles amid oily slime are simply horrible. Corralled by the circulating currents of the central North Pacific, the so-called Great Pacific Garbage

Patch may be not visible to passing mariners but its debris, especially plastic particles, menaces birds. In the mid-1990s a study of 251 dead or injured Laysan Albatross chicks on Midway Island, at the north-western end of the Hawaiian chain, discovered, depressingly, that all but six contained plastic debris.[6] Since then, the plight of dead albatross chicks with rotting bellies chock-full of plastic waste has been widely advertised.[7] "And till my ghastly tale is told, this heart within me burns" wrote Samuel Taylor Coleridge in *The Rime of the Ancient Mariner.*[8] Have the two centuries since Coleridge's writing witnessed a deterioration in our management of the seas' creatures? If human activity can wreak such distress on innocent albatross chicks so distant from urban sprawls, has humanity abrogated responsibility for stewardship of the planet?

I hope the answer to those questions is no. That said, the solution of the problems of marine pollution will rarely entail improved knowledge of the habits of seabirds derived from modern technology. Rather, the solutions will involve stronger legislation and more rigorous enforcement. However, other aspects of seabird conservation certainly will benefit from improved knowledge.

In the effort to mitigate climate change, sustainable or green energy generation will be critical. At sea, this will undoubtedly involve offshore wind farms. Associated with farms are several threats to the environment. There is the disruption associated with construction, especially if the turbine is mounted on piles, and not floating. There is the question of whether the farm's presence will deter and adversely affect (or, less likely, attract) seabirds and indeed other animals in the longer term. There may be no tidy answer to this question since GPS-tracking of Lesser Black-backed Gulls has already hinted that the amount of use birds make of wind-farm areas can vary from year to year and between males and females.[9] However the divers (loons) have emerged as a group of birds wary of offshore wind farms. Further, when the Nysted wind farm was built in the Baltic off southern Denmark, the percentage of flocks, mostly Common Eiders and geese migrating by day, entering the wind farm area decreased by a factor of 4.5 when pre-construction was compared to operation.[10]

Finally, once within the perimeter of the wind farm, birds face a significant possibility of impact. With the largest, 80 m blades of modern turbines rotating 20 times a minute, the blade's tip is moving at close to 600 km/h, manifestly a danger to flying birds. The extent of the danger will hinge on the heights at which birds fly and, if they fly at a vulnerable height, whether the birds can detect and avoid the blades. Both flight heights and avoidance behaviour are active areas of research. For example Lesser Black-backed Gulls breeding in the United Kingdom fly lower by night than by day. Fortunately, if fortuitously, this means that the gulls are flying lower when they are less likely to be able to see the Swords of Damocles or, more prosaically, the turbine blades.[11]

Northern Gannets have been studied when setting forth to collect food for their chicks from the Bass Rock colony in the Firth of Forth, Scotland. Thanks to the exertions of scientists from the Universities of Exeter and Leeds, the 55 adult gannets carried GPS trackers that revealed their height above the sea, as did the atmospheric pressure sensors carried by a subset of the birds. When commuting to feeding areas, often over sea areas with planning consent for wind farms, the median height of the gannets was 11 m above the sea. The danger from blades, whose minimum permitted height is 22 m, would be small but not negligible. However once the gannets found fish, they tended to circle and rise higher, the better to scan the water below. This behaviour took them to a median height of 26 m, well within the savage sweep of a turbine blade.[12]

Undoubtedly the risk will vary among species. Within British waters, a low-flying Little Auk is virtually always safe from impact with blades whereas a relatively high-flying Black-throated Diver is in greater peril, should it choose to fly through a wind farm.

Once inside a wind farm's boundaries, the bird can avoid a rotating blade by skirting around the turbine, or it can take emergency action at the very last moment when within 10 m of the blade, so-called micro-avoidance. Just how successful these avoidance behaviours are is being studied by high-resolution digital cameras, both visual and infra-red,* radar, and laser rangefinder technologies. The gannet project could bring no direct observations of avoidance to bear. However, the researchers assumed 99 percent avoidance, probably realistic, and predicted the death

* To allow observation by night when many seabirds remain at risk.

of 300 adult gannets a month, or 1,500 over the course of a Bass Rock breeding season, if two proposed farms were constructed.

Twice a day, tides ebb and flow over the seabed. There is the prospect of generating immense quantities of electricity via turbines anchored to the seabed, and this could be especially important in Scotland, potentially the beneficiary of one-quarter of Europe's tidal power. Conceptually the problem facing a seabird underwater is similar to that presented by wind farms to a flying bird. It can avoid the area altogether, it can bypass individual turbine arrays or it can play chicken, probably an oxymoron, and avoid a collision at the very last moment. Actually understanding the risk needs information on the underwater behaviour of auks, cormorants, and other diving species when near turbines. Critical behavioural information includes data on dive depth, duration and frequency, descent and ascent speeds, precisely the sort of information that is fast becoming available from depth recorders and accelerometers.

* * *

Fishing is likely to be most efficient wherever and whenever the target prey, be they fish or squid or swarming krill, are most concentrated. That is as true for birds as it is for human fishers. In the tropics the booby is alert for other boobies or perhaps noddies diving over a shoal of small fish seething in terror above a school of tuna. If the booby is quick it will be able to seize the moment - and the prey - before the shoal disperses. If the fisher is quick, there will be time to position a seine net in a circle around the tuna, scoop the school, and then smile broadly as the quivering fish, a ton of dollars, slide across the fishing vessel's deck into the ice.

Since fishers and birds are likely to be exploiting the same sea areas, there is every possibility of interaction at many levels. Even with an array of detection devices on the bridge, a shrewd captain is likely to watch birds as a further means of guiding his vessel towards fish. And the modern captain of a tuna purse seine boat may well be eyeing his radar to use bird echoes to guide him towards the tuna. This is exactly the practise of French captains in the Indian Ocean.

Conversely birds long ago learnt that fishing vessels were a source of fish. These fish are easily snatched; perhaps they have been discarded dead overboard or are wriggling wounded out of a trawl. Or perhaps it is the offal, the waste after the fish are gutted, that is the attraction. Certainly it is no surprise that seabirds are attracted to fishing vessels and 'free' food. What is surprising is the distance at which birds detect and then approach fishing vessels. When researchers examined changes in the flight direction of GPS-tracked Wandering Albatrosses in response to the toothfish fishing fleet operating around Iles Crozet, itself monitored via GPS-based vessel surveillance systems, they found increases in feeding behaviour only when the albatrosses were within 3 km of boats. But the albatrosses displayed clear changes in flight direction, towards vessels, at distances up to 30 km, a distance close to the theoretical maximum visual range of an albatross.[*13]

Comparable long-distance detection of fishing vessels is achieved by Northern Gannets in the Irish Sea. They are more likely to switch from commuting to foraging when within 11 km of a fishing vessel. The University of Exeter researchers found the gannets had remarkable additional abilities.[14] When the vessels visible to the gannets were travelling at fishing speeds, the birds were more likely to switch to foraging, and less likely to switch to commuting, when the vessels were trawlers as opposed to non-trawlers. Further, the gannets were more likely to switch to commuting when trawlers were steaming than when they were fishing. So the gannets could distinguish different types of vessel, and concentrate foraging at the vessel type, trawlers, where discarded fish were more likely, and they could discern whether the trawler was actively fishing, and likely to be a source of fish, or steaming and probably not worth pursuing.

Attractive as fishing vessels are, they also pose risks to birds. Birds can become entangled with nets and drown, or collide with trawl warps. Across the world's seas, longlines pose a fearful hazard. Sometimes tens of kilometres in length, the lines are baited with hundreds or thousands of hooks. They may be set in mid-water, often for tuna, or near

* This theoretical maximum range takes account of the curvature of the earth, the height of a fishing vessel, and the likely height above the sea of a Wandering Albatross.

The brutal moment when an Indian Yellow-nosed Albatross
is hooked astern of a longline fishing vessel.

the seabed to catch, for example, Patagonian Toothfish. During setting, as the line streams astern, there is a short period, only a few seconds, when each baited hook has not sunk far. It can be grabbed by a bird, perhaps an albatross, which is then hooked and pulled underwater. Hours later the bird's sodden drowned corpse comes back to the surface along with the retrieved line and the catch. A useless, unnecessary death, and a stain on humanity's stewardship of the seas.

To mitigate this problem, responsible for hundreds of thousands of bird deaths every year, various partial solutions are available. For example, setting at night is less likely to cause bird deaths simply because most species are then less active. Streamer lines, flapping in the wind astern of the vessel, frighten birds away from the danger zone in the immediate wake behind the propeller.

Another protective trick is to deliver the baited hook to an underwater depth beyond the reach of diving birds as speedily as possible. This can be achieved by using weighted lines or passing the line through an underwater setting tube. Such measures are especially important in the case of White-chinned Petrels, an abundant species of the Southern Ocean and a prominent scavenger behind fishing boats. Indeed they are the seabirds most commonly killed by Southern Hemisphere longline fisheries. These petrels are more active at night than most other species and consequently benefit little from night-time line-setting. They are also proficient divers, with the habit of bringing baited hooks to the surface. There, amid the throng of jostling birds, they may be displaced by larger albatrosses but, of course, the albatross risks paying a fatal price for its poor manners in muscling the petrel away from the baited hook.

Researchers from the FitzPatrick Institute of African Ornithology, situated on the elegant campus of the University of Cape Town, wanted to know the depth of White-chinned Petrel dives.[15] Using data from nine time-depth recorders deployed on breeding White-chinned Petrels, they found that 95 percent of dives were to depths less than 8 metres, but the deepest dives were to 16 metres. These depths may be unimpressive compared to those reached by shearwaters, auks, and penguins (Chapter 9), but the information is worrying. Best practise protocols for longlining stipulate that the line should be at least 5 metres down once it extends beyond the coverage of the streamer or 'tori' lines astern.

This is clearly not adequate to prevent White-chinned Petrels retrieving the bait, and potentially making it accessible to other non-diving species.

Slower steaming speeds during setting, more heavily weighted lines, and underwater setting tubes would all alleviate the danger. So would the use of hook pods. These devices, currently in development, enclose the fishing gear until it has sunk to a predetermined depth, at which point the pod opens, releasing the hook to begin fishing.

In general, longlining vessels are less attractive to seabirds than trawlers; the pickings are slimmer. That means longliners pose more of a risk when trawlers are not discarding offal or unwanted fish or indeed when they are not working at all. One study found the Scopoli's Shearwater casualties behind a longline vessel working in the Mediterranean were concentrated at the weekend when trawlers were in port, and scarce over the weekdays when trawlers were fishing.

With good will on the part of fishers, sound well-researched suggestions from the conservation community, and sensible accommodation between both sides, I am hopeful that traumatic deaths of seabirds, resulting from interactions with fishing operations, can indeed be reduced to very low levels. It is more difficult to be equally sanguine about a fundamental conflict. Fishers and seabirds are taking very roughly the same quantity of food out of the sea each year (Chapter 1), and both parties are likely to concentrate their activities in the most productive sea areas. While it is obviously too simple to say that one fish removed by a bird is one less fish to be landed to feed a hungry human family, and vice versa, the potential for conflict between the two parties is obvious.

* * *

How might the conundrum of the inevitable competition of people and birds for the ocean's finite harvest be resolved? While national measures are potentially useful, the majority of the world's seas are beyond national jurisdiction and, as we have seen, seabird journeys are frequently made across the unregulated vastness of the high seas.

The international nature of the high seas and the global web entwining birds and fishers was recently emphasized in the journal *Polar Biology*, published by the German company, Springer. Herein French researchers reported monitoring the Yellow-nosed Albatross colony on Amsterdam Island in the southern Indian Ocean. Their paper continues the story[16] "On December 29, 2011 . . . we observed an adult bird sitting on an empty nest, fitted with a curious leg band. The bird was captured and the band removed. The bird leg did not seem harmed by the ring, and the bird was apparently in good general condition. The band consisted of tightly sealed, soft-plastic tube, closed around the albatross leg with a clip of stainless metal. The waterproof, hollow translucid part contained a rolled paper, which we extracted. The paper was dry and in good condition, and had the following manuscript inscriptions: HASLINDO. 08 JAKARTA/15 - JULI - 2011: F.G./ABDULLAH LUTHER."

It transpired that the *Haslindo 8* was a longliner registered in Jakarta, Indonesia, and properly authorised to fish in the Indian Ocean. Presumably fisherman Abdullah Luther had caught the bird, either deliberately or accidentally, and decided it would be fun to release the bird with a message on its leg. And the message is now being repeated by a British author in a book from an American publisher. The multinational nature of this simple story spotlights the international, and therefore formidable, complexities of regulating high sea fisheries.

The intercourse between seabirds and fishing is global. Wherever fishing happens, it is likely to entail some interaction with seabirds and this overlap is only exacerbated because both parties, fishers and birds, are frequently driven to concentrate their efforts in the most productive seas. Where birds are using less productive seas, the likelihood of interactions with fisheries is lower, simply because it is not economically viable for fishers to target very scattered low density species.

It is easy to offer examples of the overlap of fishers and birds in the most productive seas. Non-breeding Black-browed Albatrosses from South Georgia and White-capped Albatrosses from New Zealand share the Benguela Current off Angola, Namibia and South Africa with a host of other seabirds, with Cape Fur Seals, and with fishers targeting anchovy and larger fish. In the North Pacific, satellite-tracked Short-tailed Albatrosses venture into the same regions of the Bering Sea as support the multi-million tonne fishery for Alaskan or Walleye Pollock which are processed, *inter alia*, into the filling of McDonald's Filet-O-Fish sandwiches.

So there is overlap between birds and people at sea at a coarse scale. Since it is unrealistic to expect the cessation of fishing for a hungry world, it is pertinent to ask, from the perspective of seabird conservation, whether this overlap is detrimental to birds. The likely answer is that there are some circumstances when the interaction is indeed detrimental and others where it is of little consequence. At that point the logical ambition of seabird conservationists should be to concentrate on identifying and, if possible, influencing those situations where seabirds do suffer from fishing operations.

Certainly there are situations where birds and fishers seem able to pursue their daily lives with minimal mutual interference. Macaroni Penguins from South Georgia spend the southern winter entirely at sea, a predicament deserving of sympathy. I would not wish to be swimming in 15-metre swells in the Southern Ocean with no prospect of stepping ashore from the 5°C water. This is also the time of year when the penguin's principal prey, krill, is swept up by commercial fishing. Norman Ratcliffe from the British Antarctic Survey led a team investigating whether this might be a problem.[17] With geolocator information from the birds enabling plotting of where they were spending their time, and drawing on year-round heart logger data which report on how fast penguin hearts are beating in winter and therefore how much food they require each day, Ratcliffe could map the distribution of penguin feeding effort across a swathe of the Scotia Sea extending some 20 degrees of longitude to the east and to the west of South Georgia. The map of penguin effort barely overlapped with the fishing effort. And both parties were taking only a small fraction, less than 1 percent, of the total krill stock. Conclusion: live and let live.

Another factor that may alleviate the interaction is more subtle. The oceans are very big places and it is perfectly possible that birds and fishing vessels may be active in the same area, coarsely measured, and yet rarely come into close and possibly harmful contact. Karine Delord and a group of French ornithologists attached satellite transmitters to 21 breeding White-chinned Petrels to document their travels when foraging away from the breeding colony on the Kerguelen archipelago situated at 50°S.[18] During incubation the petrels headed towards the Antarctic continent 10–15 degrees to the south. During chick feeding, they either remained close to Iles Kerguelen or went much further, to Antarctica, risking an encounter with longlining vessels in either location. Indeed the map of where the petrels and vessels respectively concentrated their efforts would lead one to suspect an unhappy conjunction. This was wrong, and it could be shown to be wrong because there was information about exactly where in the Southern Ocean the legally-operating vessels set their lines. Not one of 2,500 petrel positions, as determined from the satellite data, was within 15 km of an operating toothfish vessel. That birds and vessels were mostly minding their own business was confirmed by diet analysis: just 4 percent – or possibly slightly more – of the food fed to petrel chicks was sourced from the fishing fleet, for example bait fish.

While seabirds and fishers often and inevitably use the same broad areas, the Macaroni Penguin and White-chinned Petrel examples underscore how this is not necessarily a problem for seabird conservation in every circumstance. Rather than assume that seabirds and fishers are forever at daggers drawn, the way forward may be to identify where interaction between seabirds and fishing is most likely and then follow two routes towards conservation.

The first is to minimize the likelihood of traumatic deaths when seabirds meet a fishing vessel, as discussed earlier in the chapter. It may be that there is such a high risk that any meeting is likely to be traumatic. Then the only way to prevent such encounters is to forbid fishing – which may be feasible on a small spatial scale and/or for short periods. As mentioned in Chapter 5, the brood stage is a period when the Wandering Albatrosses of South Georgia are constrained by the needs of the small chick and do not wander far. During that brooding season, the region subject to fishery closures totally encompasses the area of highest use by Wandering Albatrosses, an ideal outcome from a conservation perspective.[19]

A second way forward is to map hotspots of seabird activity, and consider whether the ecological interaction between fisheries and seabirds, via their joint impact on fish stocks, is likely to be harmful to seabirds. This is a much less tractable proposition. Globally these hotspot areas may be so extensive that it is unrealistic to argue that all should be closed to fisheries in order to protect seabirds. Therefore the next logical

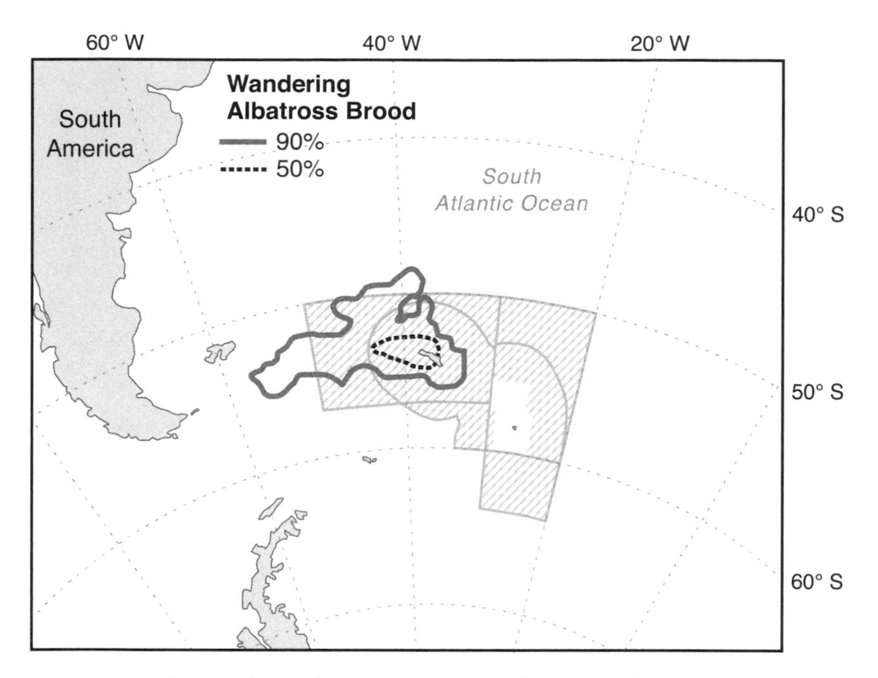


During the brood-guard stage, Wandering Albatrosses of South Georgia spend 90% of their time within the dark grey envelope and 50% within the black dashed envelope. Seasonally-applied fishery closures (cross-shading) and further protection in the 200-mile zone surrounding South Georgia and the South Sandwich Islands (grey envelope) mean the albatrosses are almost entirely protected from fishery interactions during this vulnerable stage of their life cycle. Map re-drawn from Supplementary Material attached to the work cited in Note 18, Chapter 10.

step could be to trim the list of hotspots to those where the interaction may be most harmful to seabirds, and focus conservation restrictions there, while continuing to promulgate what may be the more important message, that maintaining a healthy stock of fish (or squid or crustacea) is in the long-term interests of both birds and fishers.

Fortunately the tools to objectively map hotspots of seabird activity are becoming available. One tool, that vividly portrays the global span of seabird distribution and activity, is the Seabird Tracking Database maintained by BirdLife International.* A quick glance at the database,

* http://www.seabirdtracking.org/ (accessed 17 June 2017).

a repository of tracking data from researchers across the world, shows what probably could be guessed, that seabirds are likely to be met across the world's oceans. That observation needs refining if the aim is to identify hotspots. The process has been set in train by Ben Lascelles, for many years leader of the marine team at BirdLife International. This organisation has pioneered the identification of so-called Important Bird Areas (IBAs) on land. Designated according to rigorous criteria, these are areas where, for example, large numbers of birds, or a significant proportion (exceeding 1 percent) of a species' population, or a very rare species are found.

Transferring the IBA concept to the marine realm is not straightforward. Here are some of the problems. Given the immense tracts over which some seabird species disperse, there will be areas that are important at one time of year, only to be totally unvisited at other times. There is no guarantee that important feeding areas will not shift from year to year as ocean currents shift. The tracking data on a species are derived from a relatively small number of individuals, raising questions about how representative the tracked individuals may or may not be, and creating uncertainty about exactly how many individual birds are using what appear to be the key areas. Nevertheless the Lascelles study[20] took due account of these issues and identified 4.3 percent of the world's oceans as meeting IBA criteria, albeit with obvious gaps in, for example, the Bering and Barents Seas where there has been relatively little seabird tracking.

Four percent provides an encouraging start to the discussion. As a modest value, it avoids the risk that the seabird conservationists are perceived to be 'asking for the world', especially as protected zones already established for other reasons are likely to indirectly benefit birds. For example, along the Mediterranean coast of France where the scenery is gorgeous and the yachts swanky, three areas set aside primarily to preserve coastal fish like the Dusky Grouper and deep-sea ecosystems, have proven to be key feeding areas for the Yelkouan Shearwater.[21] Slightly further west, Critically Endangered Balearic Shearwaters that breed on Majorca head to the Catalan coast of Spain when feeding during incubation. Much of that feeding activity occurs within offshore zones already flagged as Special Protection Areas, a European Union conservation designation.[22]

The relatively modest target of the marine IBA approach is also compatible with international ambitions to increase the proportion of the world's seas in receipt of protection from the current 3 percent to 10 percent.[23] However, marine protected areas - MPAs in the jargon - are not automatically zones were fishing is banned. It may be allowed, subject to regulation.

In the laudable pursuit of seabird conservation, I wonder if the way forward may be to restrict fishing only where there is clear evidence that the activity poses a clear and present danger to seabirds. In support of that cautious approach is the argument that restricting fishing in one area can simply displace it to another area, to cause problems there. To illustrate, trawling is forbidden over most of the Patagonian Shelf between Argentina and the Falklands and, as a result, most regional interactions between seabirds, such as Black-browed Albatrosses, and fisheries occur on the border of the fishery closure area. Where the danger to seabirds is lower, then prudent conservation of the fishery stock, rather than fishery restrictions, does a favour to both birds and fishers.

None of these caveats should divert attention from the fact that there will be times and places where restrictions on fisheries benefit seabirds. In 1992 gill-net fisheries off eastern Canada for cod and Atlantic Salmon were closed. This restriction benefitted some diving seabirds, like Common Guillemots, which no longer became entangled in nets but was less favourable to other species such as gulls that ceased to enjoy the fishery discards.[24]

Or take the case of the African Penguin* whose population at the beginning of the 21st century had fallen to about 10 percent of its numbers 100 years before. As the decline continued in the early years of the 21st century, so the possibility that overfishing of anchovy, the penguins' main prey, was a cause of penguin decline gained attention. This led, from 2008 onwards, to fishery closures adjacent to penguin colonies off the South African coast. The closures extended just 20 km from the

* Also known as the Black-footed or Jackass Penguin, because of its braying call.

colonies,* in marine terms a modest area, which might benefit swimming penguins but would be of less use to flying seabirds.

These closures were coupled with studies that used GPS-tracking to assess how far afield parent penguins need to swim to collect food for their chicks whose growth rate was also measured. Possibly the adults would have to work less hard and/or the chicks would grow faster when the adjacent fishery was closed. Although the data reported to the 2nd World Seabird Conference were less crisp graphic and more smudged watercolour - this was, after all, real biology! - temporary fishery closures did appear to benefit the African Penguins. That said, the benefit to penguins was compromised by the understandable tendency of fishing operations to be concentrated just beyond the closed boundary, where very likely fish spilled out of the protected zone and rendered fishing more rewarding.

Another part of the world where there is some, but only some, evidence that seabirds have benefitted from fishery closure is the North Sea. In the 1990s, the north-western North Sea witnessed industrial fishing for sand eels, wriggling silver slivers that were transformed into fish meal for feeding livestock. This was associated with declines in the breeding success of seabirds breeding on Scotland's east coast and known from radio-tracking and satellite-tracking to be searching for sand eels for their chicks in the very fishery area. When the fishery was in full swing, the breeding success of Black-legged Kittiwakes fell, only to recover when closures were imposed in 2000. For other species, such as Razorbills and Atlantic Puffins, fishery closure provided no detectable benefit, perhaps because these species could hunt many metres underwater, in contrast to the surface-feeding Kittiwake.[25] However, the past decade has seen a continuation of the sand eel decline, not because of fishing but probably because of the sea's warming which compromises the ability of young sand eels to survive the winter while hibernating in sandy seabeds.

* An intention for the closure radius to be 20 nautical miles (nm), almost twice as great as the achieved 20 km, was lost during the passage of the necessary legislation. A cynic might notice that the proximity of k and n on a standard keyboard would increase the chance of a transcription error.

We will shortly return to the question of ocean warming and climate change. For now, the focus remains on the merits of protecting chunks of sea where seabirds concentrate. For example, in May 2014 the UK's Royal Society for the Protection of Birds (RSPB) set forth the case for seven protected areas at sea off the coast of Scotland,[26] where great seabird throngs gather in world-class colonies. Undoubtedly these offshore areas, key feeding zones for several species such as Great Skua (in fast decline) and Common Guillemot, would meet the criteria for Special Protection Areas (SPAs).* What is far less certain is whether the conservation designation would make any material difference to the well-being of the seabirds. A friend, working for the RSPB, chides my scepticism, "it's worth noting that the first thing SPAs do is to stop things getting worse. Article 6 of the Habitats Directive requires that any 'plan or project' - and fishing counts as such - proposed in a site must not have a significant impact on an SPA's [seabird] population". In other words, designating sites would allow any future plans for fishing to be squashed if the fishing posed any sort of risk to birds.

To sum up, fishing activities unquestionably cause distressing seabird deaths. Common humanity demands that these be reduced to the absolute minimum. But fishing of coastal waters and the high seas will continue for the foreseeable future. Sometimes this will have little impact on seabirds. Elsewhere, because there is an inescapable tendency for birds and fishers to gravitate towards common areas, often hotspots of seabird activity, there is the potential for interaction, notably because fishers exploit the very prey of seabirds. Since wholescale prohibition of fishing is impossible and undesirable, seabird conservation in these areas may often best be served by a focus on stock protection, in order to provide full nets for fishers and full stomachs for seabirds. Only in a carefully-identified proportion of cases is prohibition of fishing, with all the political mayhem that entails, likely to be the best solution.

* This is a designation required of European Union member states, under the EU Birds Directive, for the most important areas for birds. Frustrated by governmental inaction, the RSPB took the lead in proposing the areas. My comments ignore the considerable complications arising from the UK's imminent departure from the EU.

* * *

Climate is changing, at sea as much as on land. The North-west Passage, passing north of mainland Canada and once a paradigm of icy impenetrability, is now being visited by cruise ships. In 2016 Caspian Terns were discovered breeding in Alaska for the very first time, close to the Arctic Circle.[27] Further south in the Americas, Stephen Kress, whose working life has been a successful mission to re-establish Atlantic Puffins on the coast of Maine, was shocked in 2012. Instead of bringing cold-loving herring and hake to the chick at home in the burrow, adult Puffins started bringing a warmer water species, butterfish. The snag was that the chicks struggle to swallow this disc-shaped fish.[28] In a burrow monitored by a webcam, one luckless chick died in front of a live audience, to Kress' intense chagrin. And the problem is not local: Puffins of south-west Iceland have been faring poorly as waters have warmed. Grassy slopes in Iceland's Westman Islands, pocked with Puffin burrows, are now as bereft of birds as a turkey shed on Boxing Day. Nor are the problems confined to the North Atlantic. What will be the fate of the hapless Emperor Penguin, that necessarily breeds on sea ice, if the Antarctic ice re-distributes or even shrinks in extent? One detailed modelling study of the Emperor Penguins breeding in Terre Adélie in the French sector of Antarctica calculated a likely decrease by 2100 from the present 6,000 pairs to a meagre 400 pairs.[29] Other modelling work, encompassing all European bird species, suggests a retreat of northern seabird species: for example, Long-tailed Skuas may cease to breed along the mountain spine of Norway, and Glaucous and Ivory Gulls may abandon nesting on Svalbard.[30] And, as the icecaps melt, so the sea level will rise, a threat to coastal human communities and indeed the viability of the Maldives as a nation state, and an equal threat to the seabirds of low-lying tropical atolls.

Facing threats at sea from climate change, from fisheries and, perhaps to a lesser extent, from pollution, the prospect for seabirds in the 21st Century is by no means rosy. Addressing these issues will often not involve the technology that has generated such astounding knowledge of seabirds' seagoing lives over the last 20 years. But I can now report exceptions.

The ability to monitor tagged birds, via base stations placed in the colony to collect information on their comings and goings, and report remotely to a desk-bound researcher, will become ever more refined. I am not altogether certain that receiving a torrent of information on a computer screen will compensate for the pleasure and pain of handling an irate Puffin, with horridly sharp bill and claws, amid the slime, the smell, and the sheer life-affirming exuberance of a puffinry. However, knowledge of tagged species will certainly improve.

A changing world means colonies are likely to shift location and change in size. Such changes can be assessed in the traditional way, by painstakingly counting birds and by ground-based mapping of the area occupied. There is also the modern alternative, using drones and satellites.

One might think that drone counts would be less accurate than those obtained by experienced researchers who get up close and personal on the ground. This notion was comprehensively overturned by the recent work of Monash University biologists.[31] The team flew a $1,500 octocopter drone over three breeding colonies of Crested Terns and five of Lesser Frigatebirds on islands off the Kimberley region of north-western Australia. They also went south to sub-Antarctic Macquarie Island where a $3,000 fixed-wing drone flew above three crowds of moulting Royal Penguins. During the drones' flights, which lasted between 4 and 20 minutes, the onboard camera took photos of the birds every two to three seconds. After the images were merged, the birds were counted on computer screens – and the counts compared with those obtained on the ground by observers who visited the colonies to count the birds in the flesh at the very same time as the drone flights. Crucially the birds showed no signs of alarm as the drones flew overhead. Equally important, the counts obtained from drones showed less variation. That greater precision of the drone counts means that any changes in colony numbers, due to long-term changes in the environment, can be more reliably assessed.

The drone technique of course requires knowledge of the whereabouts of the colony. Such knowledge will not be available if, for example, Sooty Terns are shifting from an atoll swamped by sea level rise to another one, or if Emperor Penguins need to relocate their colony by tens of kilometres. However, these upheavals could be detected if seabird colonies were visible from space. They are.

Remote sensing from satellites is already able to detect seabird colonies. In 2012, Pete Fretwell and colleagues from the British Antarctic Survey published a paper that was justifiably, if a tad grandiosely, titled 'An Emperor Penguin Population Estimate: The First Global, Synoptic Survey of a Species from Space.'[32] Examination of satellite imagery of the entire Antarctic coastline had revealed four new, and hitherto unknown colonies, and confirmed the location of three previously suspected sites. This elevated the total number of known Emperor Penguin breeding colonies to 46. Arguably the more complicated part of the study was using the images to estimate the number of birds since the penguins were often tightly clustered and not distinguishable as individuals, unlike the birds in the drone photographs. To overcome this hurdle, ground-truthing was necessary. At a small number of colonies, the relationship between the number of penguins actually counted and the ground area their huddle occupied was established. This enabled the number of penguins in other clumps extending over various areas, as measured from the satellite images of the different colonies, to be calculated. And, wonderfully, a global estimate of 238,000 breeding pairs emerged. This total represented a near-doubling of the previous estimate and was celebrated worldwide from Sky News Australia to the Wall Street Journal.

With the principle proved, similar techniques have been applied to the other penguin of the far south, the Adélie Penguin. How exciting if the technique could be extended to other species, like the Antarctic Petrel, where the number of birds encountered at sea cannot be reconciled with the number at known breeding colonies, raising a strong suspicion that there are colonies yet to be discovered. Also under investigation is the idea of using satellite imagery to count the population of the rare Short-tailed Albatross on the Senkaku Islands. Because of territorial disputes between Japan and China, the islands have been off-limits for ground-based counts for over 10 years.

Another possibility is to use aerial reconnaissance to discover colonies of species whose nesting grounds remain totally unknown. One such is Hornby's Storm Petrel, mentioned earlier in the chapter. Mummified chicks have been found in the supremely arid Atacama Desert of coastal northern Chile and southern Peru. That region is almost certainly where the species breeds. Yet only one small colony has been found (see Chapter 1). Perhaps the guano imprint of colonies could be detected from the sky or from space. And, in the future, annual monitoring of the extent of tern colonies on tropical islands set in the bluest seas should be possible. How tragic it would be if the advent of these modern techniques coincided with a decline of the colonies to oblivion.

* * *

This book has attempted to paint a picture of how modern devices have enabled researchers to discover more about the lives of seabirds at sea. That simple sentence undersells the reality. Posed 50 or even 20 years ago, certain questions would have been totally unanswerable. Now, for many birds, they can be answered with some confidence. Not only can trans-oceanic flights be tracked with startling accuracy, it is also possible to tell, to within metres, where a breeding seabird is, whether it is flying or swimming and, if it is swimming, whether it is at the surface or underwater. If it is underwater, is its dive shallow or deep? It is possible to monitor the bird's heart beat and when it gulps down food. Although I admitted in Chapter 1 that it would be an exaggeration to assert that a seabird can be monitored in as much detail as a person in intensive care, or a Tour de France cyclist, it is, as we have discovered, only a small exaggeration.

Sometimes the acquisition of scientific knowledge about the natural world is reckoned to diminish wonder, as if it were a zero sum game – more knowledge necessarily equates to less wonder. From a personal perspective that position is false when applied to rainbows. Even if my knowledge of the diffraction of light of different colours is not especially deep and sophisticated, my wonder at and delight in the unwoven rainbow remains. The same applies to knowledge about seabirds.

Rockhopper penguins are porpoising towards Bleaker Island in the Falklands. The name is a corruption of Breaker Island, for the breakers are indeed massive, even frightening. I can wonder whether the penguins are porpoising to reduce their travel costs and still thrill when they scrabble ashore. Flippers beating, they struggle to grip the rocks with their clawed feet before being washed off by the next surging wave. But Rockhoppers are nothing if not persevering. They are saved from injury by a robust body that bounces like a rubber ball. Eventually the bird I am watching, a female, escapes the waves' suction and hops up the coastal cliff to relieve her mate. The relieved male will head to sea and, as we now know from satellite-tracking, travel several hundred kilometres from the colony before he returns to Bleaker.

It is a May evening on the grassy green island of North Rona. Lying some 70 km north-west of Cape Wrath, itself the north-western projection of Scotland, North Rona is the remotest British island ever to have sustained permanent settlement. The light is now ebbing from the sky above the turf-clad walls of the oratory where St Ronan possibly practised religious asceticism in the 8th century. As partial night develops – the island is too far north for total midsummer darkness – there is a vibrant elastic call. Leach's Storm Petrel! After a winter at sea, the birds are returning to their nests, small cracks in the ancient walls. Where have they spent the winter? Geolocators have identified two principal wintering areas used by the Leach's Petrels of Nova Scotia, where the species is much more numerous than in Great Britain. Those areas are tropical waters between Brazil and west Africa, and seas further south off Namibia. Perhaps the Rona petrels join their Canadian cousins in the winter; perhaps they visit another, altogether different sea area. Nobody knows. Cushioned by soft turf, I can lie on my back eyes open, and toss these questions into the deep violet night sky. The contented questioning reverie is briefly interrupted by a fleeting glimpse of a narrow-winged petrel overhead.

There is no conflict between the thrilling discoveries of the past 20 years and continuing to marvel at the ability of seabirds to cope with a salty medium so different to land. That ability will be compromised if seabirds' nesting places are rendered less secure, often by introduced alien species, and if life at sea becomes more dangerous. That will happen if the seas are over-fished, if fishing practices create perils for seabirds and if pollution is not controlled. It is not in the interests of seabirds or humanity for these changes to happen. They need not.

# ACKNOWLEDGEMENTS

This book arises from travels that have taken me to every ocean of the world, supported by the kindness of friends and strangers. There are too many to thank individually and, in some cases, I never knew their names. I think, for example, of the Chilean lorry driver who stopped to ask what I was doing on a remote and fearsomely steep mining road in the Atacama desert. When I told him that I was looking for *golondrinas de mar* (storm petrels), he directed me to a salt mine, a visit that indirectly led years later to the discovery of a new colony of a barely-known species. I think too of the Greek captain of the *Ithaca Reefer*, the tramp steamer on which I first travelled to the Pitcairn Islands. Because I never grasped his actual name, he became Captain Archimedes in my mind, and encouraged me to eat lambs' brains when the ship's company celebrated St. Nicholas' Day. I desisted.

But there are also the people and organisations that have provided longer term support. Chief among these are the Departments of Zoology at the Universities of Oxford and then Cambridge that, for 40 years, have furnished me with a home base from which to set forth to explore our wonderful world, and a desk at which to write up the results of those explorations. Colleagues at those Departments have been continually curious and supportive. This was no less true of members of the Percy FitzPatrick Insitute, University of Cape Town, through which I experienced the rumbustious Southern Ocean, the domain of albatrosses and penguins.

Many seabird enthusiasts have kindly made time to answer queries or check small sections of text. They include Mark Bolton, Maria Dias, Annette Fayet, Tony Gaston, Tim Guilford, Fiona McDuie, Børge Moe, Tamsin O'Connell, Steffen Oppel, Ollie Padget, Richard Phillips, Matt Rayner, Nils Røv, Cleo Small, Philip Taylor, Ewan Wakefield, Ross Wanless, Bryan Watts, Henri Weimerskirch, and Barbara Wienecke. A small group has been generous enough to read all - or nearly all - of the text and provide countless helpful suggestions. The group contains Tim Birkhead, Tim Brooke, Tommy Clay, Nick Davies, John Fanshawe, Adrian Friday, Terry Jones, Peter Ryan, and Eric Woehler.

Deserving of a special mention is Mark Maftei whose rap presentation electrified the 2nd World Seabird Conference in 2015. This tour-de-force caught me by surprise. Fortunately, after the presentation, Mark was generous enough to retire to a quiet corner and repeat the lyrics into a digital recorder, allowing me to transcribe the verses at leisure, and quote some them in this book.

The following photographers generously supplied photos for possible use in the book: Peter Becker, Abe Borker, Dave Boyle, Annette Fayet, Jacob González-Solís, Jamie Gundry, Oliver Krüger, Mike Harris, Bungo Nishizawa, Richard Phillips, Matt Rayner, Maties Rebassa, Sabrina Weitekamp, Laurie Wilson, Rory Wilson, and Watanuki Yutaka. Alas, the constraints of space meant not all their submissions could be used.

Bruce Pearson has an extraordinary ability to capture the immensity of the sea and the character of the creatures that dwell there in a few strokes of his brush or pen. This facility is wonderfully evident in the artwork he has supplied to enhance the book. I am indeed grateful.

At Princeton University Press, a small team worked tirelessly and constructively to turn a variety of manuscript files into a coherent book. Led by editor Ingrid Gnerlich, the team included David Campbell, Kathleen Cioffi, Frances Cooper, James Curtis, Meghan Kanabay, Dimitri Karetnikov, Julia Hall, Sara Henning-Stout, Catja Pafort, Stephanie Rojas, Pamela Schnitter, and Arthur Werneck.

Scientists, even those conducting studies on the world's most distant island specks, know full well that any success associated with their endeavours depends on building upon past knowledge, on standing on the shoulders of giants. This has been true of my own efforts, and is even more true of this book which has relied on the industry of a worldwide community of hundreds of researchers. It is always risky to single out individuals, but I am terribly conscious that this book would certainly have been the poorer and probably impossible without the research output of the teams inspired by Tim Guilford at Oxford University, Richard Phillips at the British Antarctic Survey, Henri Weimerskirch of the French Centre National de la Recherche Scientifique, and Rory Wilson of Swansea University. Thank you all.

# LIST OF BIRD SPECIES MENTIONED AND THEIR SCIENTIFIC NAMES

The birds are arranged in alphabetical order of their common names.

| | |
|---|---|
| Adélie Penguin | *Pygoscelis adeliae* |
| African Penguin | *Spheniscus demersus* |
| Ancient Murrelet | *Synthliboramphus antiquus* |
| Antarctic Petrel | *Thalassoica antarctica* |
| Antarctic Prion | *Pachyptila desolata* |
| Antarctic Tern | *Sterna vittata* |
| Arctic Tern | *Sterna paradisaea* |
| Ascension Frigatebird | *Fregata aquila* |
| Atlantic Puffin | *Fratercula arctica* |
| Atlantic Yellow-nosed Albatross | *Thalassarche chlororhynchos* |
| Balearic Shearwater | *Puffinus mauretanicus* |
| Barau's Petrel | *Pterodroma baraui* |
| Bar-tailed Godwit | *Limosa lapponica* |
| Black Petrel | *Procellaria parkinsoni* |
| Black-browed Albatross | *Thalassarche melanophrys* |
| Black-footed Albatross | *Phoebastria nigripes* |
| Black-capped Petrel | *Pterodroma hasitata* |
| Black-legged Kittiwake | *Rissa tridactyla* |
| Blackpoll Warbler | *Dendroica striata* |
| Black-throated Diver | *Gavia arctica* |
| Blue Petrel | *Halobaena caerulea* |

| | |
|---|---|
| Blue-footed Booby | *Sula nebouxii* |
| Brown Booby | *Sula leucogaster* |
| Brown Pelican | *Pelecanus occidentalis* |
| Brünnich's Guillemot | *Uria lomvia* |
| Bryan's Shearwater | *Puffinus bryani* |
| Bulwer's Petrel | *Bulweria bulwerii* |
| Cape Gannet | *Morus capensis* |
| Caspian Tern | *Hydroprogne caspia* |
| Cassin's Auklet | *Ptychoramphus aleuticus* |
| Chatham Petrel | *Pterodroma axillaris* |
| Chinese Crested Tern | *Thalasseus bernsteini* |
| Christmas Island Frigatebird | *Fregata andrewsi* |
| Common Diving Petrel | *Pelecanoides urinatrix* |
| Common Eider | *Somateria mollissima* |
| Common Guillemot | *Uria aalge* |
| Common Tern | *Sterna hirundo* |
| Cook's Petrel | *Pterodroma cookii* |
| Cory's Shearwater | *Calonectris borealis* |
| Crested Tern | *Thalasseus bergii* |
| Desertas Petrel | *Pterodroma deserta* |
| Emperor Penguin | *Aptenodytes forsteri* |
| European Shag | *Phalacrocorax aristotelis* |
| Flightless Cormorant | *Phalacrocorax harrisi* |
| Galápagos Penguin | *Spheniscus mendiculus* |
| Gentoo Penguin | *Pygoscelis papua* |
| Glaucous Gull | *Larus hyperboreus* |
| Golden-winged Warbler | *Vermivora chrysoptera* |
| Great Auk | *Pinguinus impennis* |
| Great Cormorant | *Phalacrocorax carbo* |
| Great Frigatebird | *Fregata minor* |
| Great Shearwater | *Ardenna gravis* |
| Great Skua | *Stercorarius skua* |
| Great-winged Petrel | *Pterodroma macroptera* |
| Grey-faced Petrel | *Pterodroma gouldi* |

| Grey-headed Albatross | *Thalassarche chrysostoma* |
|---|---|
| Guanay Cormorant | *Leucocarbo bougainvillii* |
| Hawaiian Petrel | *Pterodroma sandwichensis* |
| Herring Gull | *Larus argentatus* |
| Hornby's Storm-petrel | *Hydrobates hornbyi* |
| House Sparrow | *Passer domesticus* |
| Imperial Shag | *Leucocarbo atriceps* |
| Indian Yellow-nosed Albatross | *Thalassarche carteri* |
| Indigo Bunting | *Passerina cyanea* |
| Ivory Gull | *Pagophila eburnea* |
| Juan Fernández Petrel | *Pterodroma externa* |
| Kerguelen Petrel | *Aphrodroma brevirostris* |
| King Penguin | *Aptenodytes patagonicus* |
| Laysan Albatross | *Phoebastria immutabilis* |
| Leach's Storm-petrel | *Hydrobates leucorhous* |
| Least Auklet | *Aethia pusilla* |
| Least Storm-petrel | *Halocyptena microsoma* |
| Lesser Black-backed Gull | *Larus fuscus* |
| Lesser Frigatebird | *Fregata ariel* |
| Light-mantled Sooty Albatross | *Phoebetria palpebrata* |
| Little Auk | *Alle alle* |
| Little Penguin | *Eudyptula minor* |
| Long-tailed Skua | *Stercorarius longicaudus* |
| Macaroni Penguin | *Eudyptes chrysolophus* |
| MacGillivray's Petrel | *Pseudobulweria macgillivrayi* |
| Magellanic Penguin | *Spheniscus magellanicus* |
| Magnificent Frigatebird | *Fregata magnificens* |
| Manx Shearwater | *Puffinus puffinus* |
| Marbled Murrelet | *Brachyramphus marmoratus* |
| Masked Booby | *Sula dactylatra* |
| Monteiro's Storm-petrel | *Hydrobates monteiroi* |
| Murphy's Petrel | *Pterodroma ultima* |
| New Zealand Storm-petrel | *Fregetta maoriana* |
| Newell's Shearwater | *Puffinus newelli* |

| | |
|---|---|
| Northern Fulmar | *Fulmarus glacialis* |
| Northern Gannet | *Morus bassanus* |
| Northern Royal Albatross | *Diomedea sanfordi* |
| Northern Wheatear | *Oenanthe oenanthe* |
| Pincoya Storm-petrel | *Oceanites pincoyae* |
| Razorbill | *Alca torda* |
| Red-footed Booby | *Sula sula* |
| Red-legged Kittiwake | *Rissa brevirostris* |
| Red-necked Phalarope | *Phalaropus lobatus* |
| (Southern) Rockhopper Penguin | *Eudyptes chrysocome* |
| Ross's Gull | *Rhodostethia rosea* |
| Royal Penguin | *Eudyptes schlegeli* |
| Sabine's Gull | *Xema sabini* |
| Scopoli's Shearwater | *Calonectris diomedea* |
| Short-tailed Albatross | *Phoebastria albatrus* |
| Short-tailed Shearwater | *Ardenna tenuirostris* |
| Shy Albatross | *Thalassarche cauta* |
| Sooty Albatross | *Phoebetria fusca* |
| Sooty Shearwater | *Ardenna grisea* |
| Sooty Tern | *Onychoprion fuscatus* |
| South Georgia Diving Petrel | *Pelecanoides georgicus* |
| South Polar Skua | *Stercorarius maccormicki* |
| Southern Fulmar | *Fulmarus glacialoides* |
| Southern Giant Petrel | *Macronectes giganteus* |
| Spoon-billed Sandpiper | *Calidris pygmaea* |
| Thin-billed Prion | *Pachyptila belcheri* |
| Wandering Albatross | *Diomedea exulans* |
| Waved Albatross | *Phoebastria irrorata* |
| Wedge-tailed Shearwater | *Ardenna pacifica* |
| Westland Petrel | *Procellaria westlandica* |
| Whimbrel | *Numenius phaeopus* |
| White-capped Albatross | *Thalassarche steadi* |
| White-chinned Petrel | *Procellaria aequinoctialis* |
| Yelkouan Shearwater | *Puffinus yelkouan* |

# NOTES

## Chapter 1. Introduction to the World's Seabirds: Past Knowledge and New Revelations

1. While I was writing this book, a GPS study of four Herring Gulls nesting on the roofs of St Ives buildings was published, showing how each bird had its individual habits (see Chapter 7). Tracked over a single breeding season, one bird ranged across the sea up to 86 km north of St Ives and one almost as widely over the Cornish countryside, but the remaining two did not venture far. Their maximum distances from the nest were 17 and 10 km. Rock, P., Camphuysen, C.J., Shamoun-Baranes, J., Ross-Smith, V.H. and Vaughan, I.P. 2016. Results from the first GPS tracking of roof-nesting Herring Gulls *Larus argentatus* in the UK. *Ringing & Migration* **31**: 47–62.
2. https://en.wikisource.org/wiki/The_Mirror_of_the_Sea/Chapter_XXVI (accessed 28 June 2017).
3. Species totals for different groups from http://www.birdlife.org/datazone/home (accessed 14 June 2017). This website also provides a source of information on Chinese Crested Tern and MacGillivray's Petrel.
4. http://www.blueridgejournal.com/poems/ci-corm.htm (accessed 28 June 2017).
5. Newly-described species from Wikipedia and from Harrison, P., Sallaberry, M., Gaskin, C.P., Baird, K.A. *et al.* 2013. A new storm-petrel species from Chile. *Auk* **130**: 180–91.
6. Murphy, R.C. 1922. Note on the Tubinares, including records which affect the A.O.U. Check-List. *Auk* **39**: 58–65.
7. Lockley, R.M. 1942. *Shearwaters*. London: J.M. Dent.
8. Lack, D. 1968. *Ecological Adaptations for Breeding in Birds*. London: Methuen.

9. Weimerskirch, H. and Cherel, Y. 1998. Feeding ecology of short-tailed shearwaters: Breeding in Tasmania and foraging in the Antarctic? *Marine Ecology Progress Series* **167**: 261–74.
10. Imber, M.J. 1973. The food of Grey-faced Petrels (*Pterodroma macroptera gouldi* (Hutton)), with special reference to diurnal vertical migration of their prey. *Journal of Animal Ecology* **42**: 645–62.
11. Brooke, M. de L. 2004. The food consumption of the world's seabirds. *Biology Letters* S4: 24648.
12. Jouventin, P., Mougin, J.-L., Stahl, J.-C. and Weimerskirch, H. 1982. La ségrégation écologiques entre les oiseaux des îles Crozet. Données préliminaire. *Comité National Français des Recherches Antarctiques* **51**: 457–67.
13. Rayner, M.J., Gaskin, C.P., Fitzgerald, N.B., Baird K.A. *et al.* 2015. Using miniaturized radiotelemetry to discover the breeding grounds of the endangered New Zealand Storm Petrel *Fregetta maoriana. Ibis* **157**: 754–66.
14. Jouventin, P. and Weimerskirch, H. 1990. Satellite tracking of wandering albatrosses. *Nature* **343**: 746–8.
15. http://vimeopro.com/south422/animal-gps-track-animation (accessed 14 June 2017).
16. For information about the latest positional devices available, see the websites of such market-leading companies as Biotrack, Holohil, and Pathtrack.
17. Delong, R.L. 1992. Documenting migrations of northern elephant seals using day length. *Marine Mammal Science* **8**: 155–9.
18. Naito, Y., Asaga, T. and Ohyama, Y. 1990. Diving behavior of Adélie Penguins determined by time-depth recorder. *Condor* **92**: 583–6.
19. Froget, G., Butler, P.J., Woakes, A.J., Fahlman, A. *et al.* 2004. Heart rate and energetics of free-ranging king penguins (*Aptenodytes patagonicus*). *Journal of Experimental Biology* **207**: 3917–26.
20. Wilson, R.P., Cooper, J. and Plötz, J. 1992. Can we determine when marine endotherms feed? A case study with seabirds. *Journal of Experimental Biology* **167**: 267–75.
21. Plötz, J., Bornemann, H., Knust, R., Schröder, A. *et al.* 2001. Foraging behaviour of Weddell seals, and its ecological implications. *Polar Biology* **24**: 901–9.
22. Hanuise, N., Bost, C.-A., Huin, W., Auber A. *et al.* 2010. Measuring foraging activity in a deep-diving bird: comparing wiggles, oesophageal temperatures and beak-opening angles as proxies of feeding. *Journal of Experimental Biology* **213**: 3874–80.
23. Watanuki, Y., Daunt, F., Takahashi, A., Newell, M. *et al.* 2008. Microhabitat use and prey capture of a bottom-feeding top predator, the European shag,

shown by camera loggers. *Marine Ecology Progress Series* **356**: 283–93. (Cameras recorded every 15s, depth sensors every second.)

24. Votier, S.C., Bicknell, A., Cox, S.L., Scales, K.L. *et al.* 2013. A bird's eye view of discard reforms: Bird-borne cameras reveal seabird/fishery interactions. *PLoS One* **8**: e57376 Doi: 10.1371/journal.pone.0057376. (Cameras and GPS loggers recorded at 1-minute intervals.)
25. Yoda, K., Sato, K., Niizuma, Y., Kurita, M. *et al.* 1999. Precise monitoring of porpoising behaviour of Adélie penguins determined using acceleration data loggers. *Journal of Experimental Biology* **202**: 3121–6.
26. Sato K., Daunt, F., Watanuki, Y., Takahashi, A. *et al.* 2008. A new method to quantify prey acquisition in diving seabirds using wing stroke frequency. *Journal of Experimental Biology* **211**: 58–65.
27. Smyth, B. and Nebel, S. 2013. Passive Integrated Transponder (PIT) Tags in the Study of Animal Movement. *Nature Education Knowledge* **4**(3): 3.
28. David Ainley's penguin work at http://www.penguinscience.com/current_sum.php (accessed 14 June 2017).
29. Professor Dr. Peter Becker's research publications at http://ifv-vogelwarte.de/en/home-ifv/staff/prof-dr-peter-h-becker/publications-peter-becker.html (accessed 14 June 2017).
30. Wilson, R.P. and Vandenabeele, S.P. 2012. Technological innovation in archival tags used in seabird research. *Marine Ecology Progress Series* **451**: 245–62. For examples of devices collecting data on several channels see: http://wildlifecomputers.com (accessed 14 June 2017).

## Chapter 2. Taking the Plunge: Seabirds' First Journeys

1. Kooyman, G.L. and Ponganis, P.J. 2008. The initial journey of juvenile emperor penguins. *Aquatic Conservation* **17**: S37–S43.
2. Ponganis, P.J., Starke, L.N., Horning, M. and Kooyman, G.L. 1999. Development of diving capacity in Emperor Penguins. *Journal of Experimental Biology* **202**: 781–6.
3. Pütz, K., Trathan, P.N., Pedrana, J., Collins, M.A. *et al.* 2014. Post-fledging dispersal of King Penguins (*Aptenodytes patagonicus*) from two breeding sites in the South Atlantic. *PLoS One* **9**: e97164. Doi: 10.1371/journal.pone.0097164.
4. Polito M.J. and Trivelpiece, W.Z. 2008. Transition to independence and evidence of extended parental care in the gentoo penguin (*Pygoscelis papua*). *Marine Biology* **154**: 231–40.
5. Willett, G. 1915. Summer birds of Forrester Island, Alaska. *Auk* **32**: 295–305.

6. Jones, I.L., Falls, J.B. and Gaston A.J. 1987. Vocal recognition between parents and young of Ancient Murrelets, *Synthliboramphus antiquus* (Aves, Alcidae). Animal Behaviour **35**(5): 1405–15.
7. Burke, C.M., Montevecchi, W.A. and Regular, P.M. 2015. Seasonal variation in parental care drives sex-specific foraging by a monomorphic seabird. *PLoS One* **10**: e0141190. Doi: 10.1371/journal.pone.0141190.
8. Yoda, K., Kohno, H. and Naito, Y. 2004. Development of flight performance in the brown booby. *Proceedings of the Royal Society of London, B.* (Suppl.) **271**: S240–S242.
9. Weimerskirch, H., Bishop, C., Jeanniard-du-Dot, T., Prudor, A. *et al.* 2016. Frigate birds track atmospheric conditions over months-long transoceanic flights. *Science* **353**: 74–8.
10. Harris, M.P. 1970. Abnormal migration and hybridization of *Larus argentatus* and *L. fuscus* after interspecies fostering experiments. *Ibis* **112**: 488–98.
11. Lockley, R.M. 1942. *Shearwaters*. London: J.M. Dent.
12. Péron, C., and Grémillet, D. 2013. Tracking through life stages: Adult, immature and juvenile Autumn migration in a long-lived seabird. *PLoS One* **8**: e72713. Doi: 10.1371/journal.pone.0072713.
13. Weimerskirch, H, Akesson, S. and Pinaud, D. 2006. Postnatal dispersal of wandering albatrosses *Diomedea exulans*: Implications for the conservation of the species. *Journal of Avian Biology* **37**: 23–8.
14. Åkesson, S. and Weimerskirch, H. 2014. Evidence for sex-segregated ocean distributions of first-winter Wandering Albatrosses at Crozet Islands. *PLoS One* **9**: e86779. Doi: 10.1371/journal.pone.0086779.
15. Gutowsky, S. E., Tremblay, Y., Kappes, M.A., Flint, E.N. *et al.* 2014. Divergent post-breeding distribution and habitat associations of fledgling and adult Black-footed Albatrosses *Phoebastria nigripes* in the North Pacific. *Ibis* **156**: 60–72.
16. Thiers, L., Delord, K., Barbraud, C., Phillips, R.A. *et al.* 2014. Foraging zones of the two sibling species of giant petrels in the Indian Ocean throughout the annual cycle: Implication for their conservation. Marine Ecology Progress Series **499**: 233–48.
17. Grissac, S., Borger, L., Guitteaud, A. and Weimerskirch, H. 2016. Contrasting movement strategies among juvenile albatrosses and petrels. *Scientific Reports* **6**: 26103. Doi: 10.1038/srep26103.
18. Blanco, G.S. and Quintana, F. 2014. Differential use of the Argentine shelf by wintering adults and juveniles southern giant petrels, *Macronectes giganteus*, from Patagonia. *Estuarine Coastal and Shelf Science* **149**: 151–9.

19. Alderman, R., Gales, R., Hobday, A.J. and Candy, S.G. 2010. Post-fledging survival and dispersal of shy albatross from three breeding colonies in Tasmania. *Marine Ecology Progress Series* **405**: 271–85.

## Chapter 3. The Meandering Years of Immaturity

1. Hector, J.A.L., Croxall, J.P. and Follett, B.K. 1986. Reproductive endocrinology of the Wandering Albatross *Diomedea exulans* in relation to biennial breeding and deferred sexual maturity. *Ibis* **128**: 9–22.
   Hector, J.A.L., Pickering, S.P.C., Croxall, J.P. and Follett, B.K. 1990. The endocrine basis of deferred sexual maturity in the Wandering Albatross, *Diomedea exulans* L. *Functional Ecology* **4**: 59–66.
2. Hobson, K.A. 1999. Tracing origins and migration of wildlife using stable isotopes: a review. *Oecologia* **120:** 314–26.
3. Catching gannets at https://www.youtube.com/watch?v=r1ZkkX9-qwM (accessed 14 June 2017).
4. Votier, S.C., Grecian, W.J., Patrick S. and Newton J. 2011. Inter-colony movements, at-sea behaviour and foraging in an immature seabird: Results from GPS-PPT tracking, radio-tracking and stable isotope analysis. *Marine Biology* **158**: 355–62.
5. Péron, C. and Grémillet, D. 2013. Tracking through life stages: Adult, immature and juvenile autumn migration in a long-lived seabird. *PLoS One* **8**: e72713. Doi: 10.1371/journal.pone.0072713.
6. Fayet, A.L., Freeman, R., Shoji, A., Padget, O. *et al.* 2015. Lower foraging efficiency in immatures drives spatial segregation with breeding adults in a long-lived pelagic seabird. *Animal Behaviour* **110**: 79–89.
7. Riotte-Lambert L. and Weimerskirch, H. 2013. Do naive juvenile seabirds forage differently from adults? *Proceedings of the Royal Society of London, B* **280**: 20131434. Doi: 10.1098/rspb.2013.1434.
8. Haug, F.D., Paiva, V.H., Werner, A.C. and Ramos J.A. 2015. Foraging by experienced and inexperienced Cory's shearwater along a 3-year period of ameliorating foraging conditions. *Marine Biology* **162**: 649–60.
9. Weimerskirch, H., Cherel, Y., Delord, K., Jaeger, A. *et al.* 2014. Lifetime foraging patterns of the wandering albatross: Life on the move! *Journal of Experimental Marine Biology and Ecology* **450**: 68–78.
10. Jaeger, A., Goutte, A., Lecomte, V.J. Richard, P. *et al.* 2014. Age, sex, and breeding status shape a complex foraging pattern in an extremely long-lived seabird. *Ecology* **95**: 2324–33.

## Chapter 4. Adult Migrations: 20,000 Leagues over the Sea

1. Gill, R.E., Tibbitts, T.L., Douglas, D.C., Handel, C.M. *et al.* 2009. Extreme endurance flights by landbirds crossing the Pacific Ocean: ecological corridor rather than barrier? *Proceedings of the Royal Society of London, B* **276**: 447–58.
2. DeLuca, W.V., Woodworth, B.K., Rimmer, C.C., Marra, P.P. *et al.* 2015. Transoceanic migration by a 12 g songbird. *Biology Letters* **11**: UNSP 20141045.
3. Egevang, C., Stenhouse, I.J., Phillips, R.A., Petersen, A. *et al.* 2010. Tracking of Arctic terns *Sterna paradisaea* reveals longest animal migration. *Proceedings of the National Academy of Sciences USA* **107**: 2078–81.
4. Fijn, R.C., Hiemstra, D., Phillips, R.A. and van der Winden, J. 2013. Arctic Terns *Sterna paradisaea* from The Netherlands migrate record distances across three oceans to Wilkes Land, East Antarctica. Ardea **101**: 3–12.
5. Croxall, J.P., Silk, J.R.D., Phillips, R.A., Afanasyev, V. *et al.* 2005. Global circumnavigations: Tracking year-round ranges of non-breeding albatrosses *Science* **307**: 249–50.
6. Guilford, T., Freeman, R., Boyle, D., Dean, B. *et al.* 2011. A dispersive migration in the Atlantic Puffin and its implications for migratory navigation. *PLoS One* **6**: e21336. Doi: 10.1371/journal.pone.0021336.
7. Jessopp, M.J., Cronin, M., Doyle, T.K., Wilson, M. *et al.* 2013. Transatlantic migration by post-breeding puffins: A strategy to exploit a temporarily abundant food resource? *Marine Biology* **160**: 2755–62.
8. Maftei, M., Davis, S.E., and Mallory, M.L. 2015. Confirmation of a wintering ground of Ross's Gull *Rhodostethia rosea* in the northern Labrador Sea. *Ibis* **152**: 642–47.
9. http://www.bou.org.uk/red-necked-phalarope-pacific-ocean/ (accessed 14 June 2017).
10. Quillfeldt, P., Cherel, Y., Masello, J.F., Delord, K., *et al.* 2015. Half a world apart? Overlap in nonbreeding distributions of Atlantic and Indian Ocean Thin-Billed Prions. *PLoS One* **10**: e0125007. Doi: 10.1371/journal.pone.0125007.
11. Spencer, N.C., Gilchrist, H.G. and Mallory, M.L. 2014. Annual movement patterns of endangered Ivory Gulls: The importance of sea ice. *PLoS One* **9**: e115231. Doi: 10.1371/journal.pone.0115231.
12. Gilg, O., Strøm, H., Aebischer, A., Gavrilo, M.V. *et al.* 2010. Post-breeding movements of northeast Atlantic Ivory Gull *Pagophila eburnea* populations. *Journal of Avian Biology* **41**: 532–42.

13. Frederiksen, M., Moe, B., Daunt, F., Phillips, R.A. *et al.* 2012. Multicolony tracking reveals the winter distribution of a pelagic seabird on an ocean basin scale. *Diversity and Distributions* **18**: 530–42.
14. Rayner, M.J., Hauber, M.E., Steeves, T.E., Lawrence, H.A. *et al.* 2011. Contemporary and historical separation of transequatorial migration between genetically distinct seabird populations. *Nature Communications* **2**: Article 332. Doi: 10.1038/ncomms1330.
15. Weimerskirch, H., Tarroux, A., Chastel, O., Delord, K. *et al.* 2015. Population-specific wintering distributions of adult south polar skuas over three oceans. *Marine Ecology Progress Series* **538**: 229–37.
16. Kopp, M., Peter, H.-U., Mustafa, O., Lisovski, S. *et al.* 2011. South Polar Skuas from a single breeding population overwinter in different oceans though show similar migration patterns. *Marine Ecology Progress Series* **435**: 263–7.
17. Gonzalez-Solis, J., Croxall, J.P., Oro, D. and Ruiz, X. 2007. Trans-equatorial migration and mixing in the wintering areas of a pelagic seabird. *Frontiers in Ecology and the Environment* **5**: 297–301.
18. Dias, M.P., Granadeiro, J.P., Phillips, R.A., Alonso, H. *et al.* 2011. Breaking the routine: Individual Cory's shearwaters shift winter destinations between hemispheres and across ocean basins. *Proceedings of the Royal Society of London, B* **278**: 1786–93.
19. Landers, T.J., Rayner, M.J., Phillips, R.A. and Hauber, M.E. 2011. Dynamics of seasonal movements by a trans-Pacific migrant, the Westland Petrel *Procellaria westlandica. Condor* **113**: 71–9.
20. Puetz, K., Schiavini, A., Rey, A.R. and Luthi, B.H. 2007. Winter migration of Magellanic Penguins (*Spheniscus magellanicus*) from the southernmost distributional range. *Marine Biology* **152**: 1227–35.
21. McKnight, A., Allyn, A.J., Duffy, D.C. and Irons, D.B. 2013. 'Stepping stone' pattern in Pacific Arctic Tern migration reveals the importance of upwelling areas. *Marine Ecology Progress Series* **491**: 253–64.
22. Duffy, D.C., McKnight, A. and Irons, D.B. 2013. Trans-Andean passage of migrating Arctic Terns over Patagonia. *Marine Ornithology* **41**: 155–9.
23. Guilford, T., Meade, J., Willis, J., Phillips, R.A. *et al.* 2009. Migration and stopover in a small pelagic seabird, the Manx shearwater *Puffinus puffinus*: Insights from machine learning. *Proceedings of the Royal Society of London, B* **276**: 1215–23.
24. Freeman, R., Dean, B., Kirk, H., Leonard, K. *et al.* 2013. Predictive ethoinformatics reveals the complex migratory behaviour of a pelagic seabird, the Manx Shearwater. *Journal of the Royal Society Interface* **10**: 20130279. Doi: 10.1098/rsif.2013.0279.

25. Stenhouse, I.J., Egevang, C. and Phillips, R.A. 2012. Trans-equatorial migration, staging sites and wintering area of Sabine's Gulls *Larus sabini* in the Atlantic Ocean. *Ibis* **154**: 42–51.
26. Gilg, O., Moe, B., Hanssen, S.A., Schmidt, N.M. *et al.* 2013. Trans-equatorial migration routes, staging sites and wintering areas of a high-Arctic avian predator: the Long-tailed Skua (*Stercorarius longicaudus*) *PLoS One* **8**: e64614. Doi: 10.1371/journal.pone.0064614.
27. Müller, M.S., Massa, B., Phillips, R.A. and Dell'Omo. G. 2015. Seabirds mated for life migrate separately to the same places: behavioural coordination or shared proximate causes? *Animal Behaviour* **102**: 267–76.
28. Dias, M.P., Granadeiro, J.P. and Catry, P. 2012. Do seabirds differ from other migrants in their travel arrangements? On route strategies of Cory's Shearwater during its trans-Equatorial journey *PLoS One* **7**: e49376. Doi: 10.1371/journal.pone.0049376.
29. Perez, C., Granadeiro, J.P., Dias, M.P., Alonso, H. *et al.* 2014. When males are more inclined to stay at home: Insights into the partial migration of a pelagic seabird provided by geolocators and isotopes. *Behavioral Ecology* **25**: 313–19.
30. Guilford, T., Wynn, R., McMinn, M., Rodriguez, A. *et al.* 2012. Geolocators reveal migration and pre-breeding behaviour of the Critically Endangered Balearic Shearwater *Puffinus mauretanicus*. *PLoS One* **7**: e33753. Doi: 10.1371/journal.pone.0033753.
31. Brown, C.R. 1985. Energetic cost of molt in Macaroni Penguins (*Eudyptes chrysolophus*) and Rockhopper Penguins (*Eudyptes chrysocome*). *Journal of Comparative Physiology B* **155**: 515–20.
32. Tanton, J.L., Reid, K., Croxall, J.P. and Trathan, P.N. 2004. Winter distribution and behaviour of Gentoo Penguins *Pygoscelis papua* at South Georgia. *Polar Biology* **27**: 299–303.
33. Weimerskirch, H., Tarroux, A., Chastel, O., Delord, K. *et al.* 2015. Population-specific wintering distributions of adult south polar skuas over three oceans. *Marine Ecology Progress Series* **538**: 229–37.
34. Cherel. Y., Quillfeldt, P., Delord, K. and Weimerskirch, H. 2016. Combination of at-sea activity, geolocation and feather stable isotopes documents where and when seabirds molt. *Frontiers in Ecology and Evolution* **4**:3. Doi: 10.3389/fevo.2016.00003.
35. Gutowsky, S.E., Gutowsky, L.F.G., Jonsen, I.D., Leonard, M.L. *et al.* 2014. Daily activity budgets reveal a quasi-flightless stage during non-breeding in Hawaiian albatrosses. *Movement Ecology* **2**: 23. Doi: 10.1186/s40462-014-0023-4.

### A Navigational Diversion


1. Emlen, S.T. 1970. Celestial rotation: its importance in the development of migratory orientation. *Science* **170**: 1198–201.
2. Wiltschko, R. and Wiltschko, W. 2009. Avian navigation. *Auk* **126**: 717–43.
3. Gagliardo, A. 2013. Forty years of olfactory navigation in birds. *Journal of Experimental Biology* **216**: 2165–71.
4. Wikelski, M., Arriero, E., Gagliardo, A., Holland, R.A. *et al.* 2015. True navigation in migrating gulls requires intact olfactory nerves. *Scientific Reports* **5**: 17061. Doi: 10.1038/srep17061.
   Gagliardo, A., Bried, J., Lambardi, P., Luschi, P. *et al.* 2013. Oceanic navigation in Cory's shearwaters: Evidence for a crucial role of olfactory cues for homing after displacement. *Journal of Experimental Biology* **216**: 2798–805.
5. Biro, D., Meade, J. and Guilford, T. 2004. Familiar route loyalty implies visual pilotage in the homing pigeon. *Proceedings of the National Academy of Sciences USA* **101**: 17440–3.
6. Benhamou, S., Bonadonna, F. and Jouventin, P. 2003. Successful homing of magnet-carrying white-chinned petrels released in the open sea. *Animal Behaviour* **65**: 729–34.
   Bonadonna, F., Chamaille-Jammes, S., Pinaud, D. and Weimerskirch, H. 2003. Magnetic cues: are they important in Black-browed Albatross *Diomedea melanophris* orientation? *Ibis* **145**: 152–5.
   Bonadonna, F., Bajzak, C., Benhamou, S., Igloi, K. *et al.* 2005. Orientation in the wandering albatross: Interfering with magnetic perception does not affect orientation performance. *Proceedings of the Royal Society of London, B* **272**: 489–95.
7. Pollonara, E., Luschi, P., Guilford, T., Wikelski, M. *et al.* 2015. Olfaction and topography, but not magnetic cues, control navigation in a pelagic seabird: displacements with shearwaters in the Mediterranean Sea. *Scientific Reports* **5**: 16486. Doi: 10.1038/srep16486.


### Chapter 5. Tied to Home: Adult Movements during the Breeding Season


1. Harris, M.P. and Wanless, S. 2016. The use of webcams to monitor the prolonged autumn attendance of Guillemots on the Isle of May in 2015. *Scottish Birds* **36**: 3–9.
2. Gonzalez-Solis, J., Becker, P.H. and Wendeln, H. 1999. Divorce and asynchronous arrival in common terns, *Sterna hirundo*. *Animal Behaviour* **58**: 1123–9.

3. Guilford, T., Meade, J., Willis, J., Phillips, R.A. *et al.* 2009. Migration and stopover in a small pelagic seabird, the Manx shearwater *Puffinus puffinus*: Insights from machine learning. *Proceedings of the Royal Society of London, B* **276**: 1215–23.
4. http://seabirdtracking.org/mapper/index.php (accessed 14 June 2017).
5. Rayner, M.J., Taylor, G.A., Gummer, H.D., Phillips, R.A. *et al.* 2012. The breeding cycle, year-round distribution and activity patterns of the endangered Chatham Petrel (*Pterodroma axillaris). Emu* **112**: 107–16. Doi: 10.1071/MU11066.
6. Pinet, P., Jaquemet, S., Phillips, R.A. and Le Corre, M. 2012. Sex-specific foraging strategies throughout the breeding season in a tropical, sexually monomorphic small petrel. *Animal Behaviour* **83**: 979–89.
7. Ashmole, N.P. 1963. The regulation of numbers of tropical oceanic birds. *Ibis* **103**b: 458–73.
8. Gaston, A.J., Elliott, K.H., Ropert-Coudert, Y., Kato, A. *et al.* 2011. Modeling foraging range for breeding colonies of Thick-billed Murres *Uria lomvia* in the Eastern Canadian Arctic and potential overlap with industrial development. *Biological Conservation* **168**: 134–43.
9. Shoji, A., Owen, E., Bolton, M., Dean, B. *et al.* 2014. Flexible foraging strategies in a diving seabird with high flight cost. *Marine Biology* **161**: 2121–9.
10. Nelson, J.B. 1978. *The Sulidae: Gannets and Boobies*. Oxford: Oxford University Press.
11. Pollet, I.L., Ronconi, R.A., Jonsen, I.D., Leonard, M.L. *et al.* 2014. Foraging movements of Leach's storm-petrels *Oceanodroma leucorhoa* during incubation. *Journal of Avian Biology* **45**: 305–14.
12. Phalan, B., Phillips, R.A., Silk, J.R.D., Afanasyev, V. *et al.* 2007. Foraging behaviour of four albatross species by night and day. *Marine Ecology Progress Series* **340**: 271–86.
13. Ronconi, R.A., Ryan, P.G. and Ropert-Coudert, Y. 2010. Diving of Great Shearwaters (*Puffinus gravis*) in cold and warm water regions of the South Atlantic Ocean. *PLoS One* **5**: e15508. Doi: 10.1371/journal.pone.0015508.
14. Edwards, E.W.J., Quinn, L.R., Wakefield, E.D., Miller, P.I. *et al.* 2013. Tracking a northern fulmar from a Scottish nesting site to the Charlie-Gibbs Fracture Zone: Evidence of linkage between coastal breeding seabirds and Mid-Atlantic Ridge feeding sites. *Deep-Sea Research* II **98**: 438–44.
15. Putz, K., Smith, J.G., Ingham, R.J. and Luthi, B.H. 2003. Satellite tracking of male Rockhopper Penguins *Eudyptes chrysocome* during the incubation period at the Falkland Islands. *Journal of Avian Biology* **34**: 139–44.

16. Ludynia, K., Dehnhard, N., Poisbleau, M., Demongin, L.M. *et al.* 2013. Sexual segregation in rockhopper penguins during incubation. *Animal Behaviour* **85**: 255–67.
17. Kirkwood, R. and Robertson, G. 1997. The foraging ecology of female Emperor Penguins in winter. *Ecological Monographs* **67**: 155–76.
18. Brooke, M. de L. 1981. How an adult Wheatear (*Oenanthe oenanthe*) uses its territory when feeding nestlings. *Journal of Animal Ecology* **50**: 683–96.
19. Phillips, R.A., Silk, J.R.D., Phalan, B., Catry, P. *et al.* 2004. Seasonal sexual segregation in two *Thalassarche* albatross species: competitive exclusion, reproductive role specialization or foraging niche divergence? *Proceedings of the Royal Society of London, B* **271**: 1283–91.
20. Shaffer, S.A., Weimerskirch, H., Scott, D., Pinaud, D. *et al.* 2009. Spatio-temporal habitat use by breeding Sooty Shearwaters *Puffinus griseus*. *Marine Ecology Progress Series* **391**: 209–20.
21. Catard A., Weimerskirch, H. and Cherel, Y. 2000. Exploitation of distant Antarctic waters and close shelf-break waters by White-chinned Petrels rearing chicks. *Marine Ecology Progress Series* **194**: 249–61.
22. Weimerskirch, H., Le Corre, M., Gadenne, H., Pinaud, D. *et al.* 2009. Relationship between reversed sexual dimorphism, breeding investment and foraging ecology in a pelagic seabird, the masked booby. *Oecologia* **161**: 637–49.
    Weimerskirch, H., Shaffer, S.A., Tremblay, Y., Costa, D.P. *et al.* 2009. Species- and sex-specific differences in foraging behaviour and foraging zones in Blue-footed and Brown Boobies in the Gulf of California. *Marine Ecology Progress Series* **391**: 267–78.
23. Harris, S., Raya Rey, A., Phillips, R.A. and Quintana, F. 2013. Sexual segregation in timing of foraging by Imperial Shags (*Phalacrocorax atriceps*): Is it always ladies first? *Marine Biology* **160**: 1249–58.
24. Hennicke, J.C., James D.J. and Weimerskirch, H. 2015. Sex-specific habitat utilization and differential breeding investments in Christmas Island Frigatebirds throughout the breeding cycle. *PLoS One* **10**: e0129437. Doi: 10.1371/journal.pone.0129437.
25. Robertson, G.S., Bolton, M., Grecian, W.J. and Monaghan, P. 2014. Inter- and intra-year variation in foraging areas of breeding kittiwakes (*Rissa tridactyla*). *Marine Biology* **161**: 1973–86.
26. Boersma, P.D., Rebstock, G.A., Frere, E. and Moore, S.E. 2009. Following the fish: Penguins and productivity in the South Atlantic. *Ecological Monographs* **79**: 59–76.

27. Sala, J.E., Wilson, R.P., Frere, E. and Quintana, F. 2012. Foraging effort in Magellanic penguins in coastal Patagonia, Argentina. *Marine Ecology Progress Series* **464**: 273–87.
28. Oppel, S., Beard, A., Fox, D., Mackley, E. *et al.* 2015. Foraging distribution of a tropical seabird supports Ashmole's hypothesis of population regulation. *Behavioral Ecology and Sociobiology* **69**: 915–26.
29. Furness, R.W. and Birkhead, T.R. 1984. Seabird colony distributions suggest competition for food supplies during the breeding season. *Nature* **31**: 655–6.
30. Wakefield, E.D., Bodey, T.W., Bearhop, S., Blackburn, J. *et al.* 2013. Space partitioning without territoriality in Gannets. *Science* **341**: 68–70.
31. Dean, B., Freeman, R., Kirk, H., Leonard, K. *et al.* 2013. Behavioural mapping of a pelagic seabird: Combining multiple sensors and a hidden Markov model reveals the distribution of at-sea behaviour. *Journal of the Royal Society Interface* **10**: 20120570. Doi: 10.1098/rsif.2012.0570.
32. Evans, J.C., Dall, S.R.X., Bolton, M., Owen, E. *et al.* 2016. Social foraging European shags: GPS tracking reveals birds from neighbouring colonies have shared foraging grounds. *Journal of Ornithology* **157**: 23–32.
33. Whitehead, T.O., Kato, A., Ropert-Coudert, Y. and Ryan P.G. 2016. Habitat use and diving behaviour of Macaroni *Eudyptes chrysolophus* and Eastern Rockhopper *E. chrysocome filholi* Penguins during the critical pre-moult period. *Marine Biology* **163**: 19. Doi: 10.1007/s00227-015-2794-6.
34. Linnebjerg, J.F., Fort, J., Guilford, T., Reuleaux, A. *et al.* 2013. Sympatric breeding auks shift between dietary and spatial resource partitioning across the annual cycle. *PLoS One* **8**: e72987. Doi: 10.1371/journal.pone.0072987.

## Chapter 6. Wind and Waves: Friend and Foe

1. Catry, P., Phillips, R.A. and Croxall, J.P. 2004. Sustained fast travel by a Gray-headed Albatross (*Thalassarche chrysostoma*) riding an Antarctic storm. *Auk* **121**: 1208–13.
2. http://yoda-ken.sakura.ne.jp/yoda_lab/English.html (accessed 14 June 2017).
3. Cotte, C., Park, Y.-H., Guinet, C. and Bost, C.-A. 2007. Movements of foraging King Penguins through marine mesoscale eddies. *Proceedings of the Royal Society of London, B* **274**: 2385–91.
4. Weimerskirch, H., Guionnet, T., Martin, J., Shaffer, S.A. *et al.* 2000. Fast and fuel efficient? Optimal use of wind by flying albatrosses. *Proceedings of the Royal Society of London, B* **267**: 1869–74.

5. Weimerskirch, H., Le Corre, M., Ropert-Coudert, Y., Kato, A. *et al.* 2005. The three-dimensional flight of Red-footed Boobies: Adaptations to foraging in a tropical environment? *Proceedings of the Royal Society of London, B* **272**: 53–61.
6. Ropert-Coudert, Y., Wilson, R.P., Grémillet, D., Kato, A. *et al.* 2006. Electrocardiogram recordings in free-ranging gannets reveal minimum difference in heart rate during flapping versus gliding flight. *Marine Ecology Progress Series* **328**: 275–84.
7. Weimerskirch, H., Chastel, O., Barbraud, C. and Tostain, O. 2003. Flight performance: Frigatebirds ride high on thermals. *Nature* **421**: 333–34.
8. Rattenborg, N.C., Voirin, B., Cruz, S.M., Tisdale, R. *et al.* 2016. Evidence that birds sleep in mid-flight. *Nature Communications* **7**: 12468. Doi: 10.1038/ncomms12468.
9. Fritz, H., Said, S. and Weimerskirch, H. 2003. Scale-dependent hierarchical adjustments of movement patterns in a long-range foraging seabird. *Proceedings of the Royal Society of London, B* **270**: 1143–8.
10. Guilford, T.C., Meade, J., Freeman, R., Biro, D. *et al.* 2008. GPS tracking of the foraging movements of Manx Shearwaters *Puffinus puffinus* breeding on Skomer Island, Wales. *Ibis* **150**, 462–73.
11. Paiva, V.H., Guilford, T., Meade, J., Geraldes, P. *et al.* 2010. Flight dynamics of Cory's Shearwater foraging in a coastal environment. *Zoology* **113**: 47–56.
12. Fritz, H., Said, S. and Weimerskirch, H. 2003. Scale-dependent hierarchical adjustments of movement patterns in a long-range foraging seabird. *Proceedings of the Royal Society of London, B* **270**: 1143–8.
13. Amélineau, F., Péron, C., Lescroël, A., Authier, M. *et al.* 2014. Windscape and tortuosity shape the flight costs of Northern Gannets. *Journal of Experimental Biology* **217**: 876–85. Doi: 10.1242/jeb.097915.
14. Weimerskirch, H., Guionnet, T., Martin, J., Shaffer, S.A. *et al.* 2000. Fast and fuel efficient? Optimal use of wind by flying albatrosses. *Proceedings of the Royal Society of London, B* **267**: 1869–74.
15. Murray, M.D., Nicholls, D.G., Butcher, E. and Moors, P.J. 2003. How Wandering Albatrosses use weather systems to fly long distances. 2. The use of eastward-moving cold fronts from Antarctic LOWs to travel westwards across the Indian Ocean. *Emu* **103**: 59–65.
16. Weimerskirch, H., Louzao, M., de Grissac, S. and Delord, K. 2012. Changes in wind pattern alter albatross distribution and life-history traits. *Science* **335**: 211–14.
17. Hedd, A., Montevecchi, W.A., Otley, H., Phillips, R.A. *et al.* 2012. Trans-equatorial migration and habitat use by Sooty Shearwaters *Puffinus griseus*

from the South Atlantic during the nonbreeding season. *Marine Ecology Progress Series* 449: 277–90.

18. Fijn, R.C., Hiemstra, D., Phillips, R.A. and van der Winden, J. 2013. Arctic Terns *Sterna paradisaea* from The Netherlands migrate record distances across three oceans to Wilkes Land, East Antarctica. Ardea **101**: 3–12.
19. Carey, M.J., Phillips, R.A., Silk, J.R.D. and Shaffer, S.A. 2014. Trans-equatorial migration of Short-tailed Shearwaters revealed by geolocators. *Emu* **114**: 352–9.
20. Felicisimo, Á.M. Muñoz, J. and Gonzalez-Solis, J. 2008. Ocean surface winds drive dynamics of transoceanic aerial movements. *PLoS One* **3**: e2928. Doi: 10.1371/journal.pone.0002928.
21. Lewis, S., Phillips, R.A., Burthe, S.J., Wanless, S. *et al.* 2015. Contrasting responses of male and female foraging effort to year-round wind conditions. *Journal of Animal Ecology* **84**: 1490–6.
22. Ropert-Coudert, Y., Kato, A., Meyer, X., Pellé, M. *et al.* 2015. A complete breeding failure in an Adélie penguin colony correlates with unusual and extreme environmental events. *Ecography* **38**: 111–13.
23. https://birdsnews.com/2011/first-surviving-broad-billed-prions-being-released-after-massive-july-'wreck'-in-new-zealand/#.WULh1MYw1E7 (accessed 14 June 2017).
24. Fisher, J. and Lockley, R.M. 1954. *Seabirds*. London: Collins
25. James, D.A., Smith, K.G., Neal, J.C. and Hehr, J.G. 2010. A cargo of birds to Arkansas, the hurricanes in 2008 and the swept clean hypothesis. *Journal of the Arkansas Academy of Science* **64**: 86–91.
26. Nisbet, I.C.T., Wingate, D.B. and Szczys, P. 2010. Demographic consequences of a catastrophic event in the isolated population of Common Terns at Bermuda. *Waterbirds* **33**: 405–10.
27. Streby, H.M., Kramer, G.R., Peterson, S.M., Lehman, J.A. *et al.* 2015. Tornadic storm avoidance behavior in breeding songbirds. *Current Biology* **25**: 98–102.
28. http://www.nytimes.com/2012/11/13/science/birds-have-natural-ability-to-survive-storms.html (accessed 14 June 2017).

## Chapter 7. Stick or Twist? The Consistent Habits of Individuals

1. Hamer, K.C., Phillips, R.A., Hill, J.K., Wanless, S. *et al.* 2001. Contrasting foraging strategies of Gannets *Morus bassanus* at two North Atlantic colonies: Foraging trip duration and foraging area fidelity. *Marine Ecology Progress Series* **224**: 283–90.

2. Wakefield, E.D., Cleasby, I.R., Bearhop, S., Bodey, T.W. *et al.* 2015. Long-term individual foraging site fidelity – why some gannets don't change their spots. *Ecology* **96:** 3058–74.
3. Patrick, S.C., Bearhop, S., Bodey, T.W., Grecian, W.J. *et al.* 2015. Individual seabirds show consistent foraging strategies in response to predictable fisheries discards. *Journal of Avian Biology* **46**: 431–40.
4. Bayliss, A.M.M., Orben, R.A., Pistorius, P., Brickle, P. *et al.* 2015. Winter foraging site fidelity of King Penguins breeding at the Falkland Islands. *Marine Biology* **162**: 99–110.
5. Weimerskirch, H., Delord, K., Guitteaud, A., Phillips, R.A. *et al.* 2015. Extreme variation in migration strategies between and within wandering albatross populations during their sabbatical year, and their fitness consequences. *Scientific Reports* 5: 8853. Doi: 10.1038/srep08853.
6. Ramirez, I., Paiva, V.H., Fagundes, I., Menezes, D. *et al.* 2016. Conservation implications of consistent foraging and trophic ecology in a rare petrel species: Ecological consistency of a rare petrel species. *Animal Conservation* **19**: 139–52.
7. Phillips, R.A., Silk, J.R.D., Croxall, J.P., Afanasyev, V. *et al.* 2005. Summer distribution and migration of nonbreeding albatrosses: Individual consistencies and implications for conservation. *Ecology* **86**: 2386–2396.
8. Grist, H., Daunt, F., Wanless, S., Nelson, E.J. *et al.* 2014. Site fidelity and individual variation in winter location in partially migratory European Shags. *PLoS One* **9**: e98562. Doi: 10.1371/journal.pone.0098562.
9. Guilford, T., Freeman, R., Boyle, D., Dean, B. *et al.* 2011. A dispersive migration in the Atlantic Puffin and its implications for migratory navigation. *PLoS One* **6**: e21336. Doi: 10.1371/journal.pone.0021336.
10. Fayet, A. L., Freeman, R., Shoji, A., Boyle, D. *et al.* 2016. Drivers and fitness consequences of dispersive migration in a pelagic seabird. *Behavioral Ecology* **27**: 1061–72.
11. McFarlane Tranquilla, L.A., Montevecchi, W.A., Fifield, D.A., Hedd, A. *et al.* 2014. Individual winter movement strategies in two species of murre (*Uria* spp.) in the Northwest Atlantic. *PLoS One* **9**: e90583. Doi: 10.1371/journal.pone.0090583.
12. Cleasby, I.R., Wakefield, E.D., Bodey, T.W., Davies, R.D. *et al.* 2015. Sexual segregation in a wide-ranging marine predator is a consequence of habitat selection. *Marine Ecology Progress Series* **518**: 1–12.
13. Harris, M.P., Daunt, F., Bogdanova, M.I., Lahoz-Monfort, J.J. *et al.* 2013. Inter-year differences in survival of Atlantic puffins *Fratercula arctica* are not associated with winter distribution. *Marine Biology* **160**: 2877–89.

14. Patrick, S.C. and Weimerskirch, H. 2014. Consistency pays: Sex differences and fitness consequences of behavioural specialization in a wide-ranging seabird. *Biology Letters* **10**: 20140630. Doi: 10.1098/rsbl.2014.0630.

## Chapter 8. Where Seabirds Find Food

1. Burger, A.E. 1997. Arrival and departure behavior of Common Murres at colonies: Evidence for an information halo? *Colonial Waterbirds* **20**: 55–65.
2. Weimerskirch, H., Bertrand, S., Silva, J., Marques, J.C. *et al.* 2010. Use of social information in seabirds: Compass rafts indicate the heading of food patches. *PLoS One* **5**: e9928. Doi: 10.1371/journal.pone.0009928.
3. Ballance, L.T., Pitman, R.L. and Reilly, S.B. 1997. Seabird community structure along a productivity gradient: Importance of competition and energetic constraint. *Ecology* **78**: 1502–18.
4. Nevitt, G.A., Losekoot, M. and Weimerskirch, H. 2008. Evidence for olfactory search in Wandering Albatross, *Diomedea exulans*. *Proceedings of the National Academy of Sciences USA* **105**: 4576–81.
5. Grémillet, D., Fort, J., Amélineau, F., Zakharova, E. *et al.* 2015. Arctic warming: Nonlinear impacts of sea-ice and glacier melt on seabird foraging. *Global Change Biology* **21**: 1116–23.
6. Bost, C.-A., Charrassin, J.B., Clerquin, Y., Ropert-Coudert, Y. *et al.* 2004. Exploitation of distant marginal ice zones by king penguins during winter. *Marine Ecology Progress Series* **283**: 293–7.
7. Wakefield, E.D., Phillips, R.A. and Belchier, M. 2012. Foraging black-browed albatrosses target waters overlaying moraine banks - a consequence of upward benthic-pelagic coupling? *Antarctic Science* **24**: 269–80.
8. Freeman, R., Todd, D., Landers T., Thompson, D. *et al.* 2010. Black Petrels (*Procellaria parkinsoni*) patrol the ocean shelf-break: GPS tracking of a vulnerable procellariiform seabird. *PLoS One* **5**: e9236. Doi: 10.1371/journal.pone.0009236.
9. Imber, M.J. 1999. Diet and feeding ecology of the Royal Albatross *Diomedea epomophora* - King of the shelf break and inner slope. *Emu* **99**: 200–11.
10. Waugh, S., Filippi, D., Fukuda, A., Suzuki, M. *et al.* 2005. Foraging of royal albatrosses, *Diomedea epomophora*, from the Otago Peninsula and its relationships to fisheries. *Canadian Journal of Fisheries and Aquatic Sciences* **62**: 1410–21.
11. Dean, B., Kirk, H., Fayet, A., Freeman, R. *et al.* 2015. Simultaneous multi-colony tracking of a pelagic seabird reveals cross-colony utilization of a shared foraging area. *Marine Ecology Progress Series* **538**: 239–48.

12. Shaffer, S.A., Weimerskirch, H., Scott, D., Pinaud, D. *et al.* 2009. Spatiotemporal habitat use by breeding sooty shearwaters *Puffinus griseus. Marine Ecology Progress Series* **391**: 209–20.
13. Charrassin, J. and Bost, C.-A. 2001. Utilisation of the oceanic habitat by King Penguins over the annual cycle. *Marine Ecology Progress Series* **221**: 285–97.
14. Cotté, C., Park, Y.-H., Guinet, C. and Bost C.-A. 2007. Movements of foraging King Penguins through marine mesoscale eddies. *Proceedings of the Royal Society of London, B* **274**: 2385–91.
15. Ramírez, I., Paiva, V.H., Menezes, D., Silva, I. *et al.* 2013. Year-round distribution and habitat preferences of the Bugio petrel. *Marine Ecology Progress Series* **476**: 269–84.
16. Weimerskirch, H., Le Corre, M., Jaquemet, S. and Marsac, F. 2005. Foraging strategy of a tropical seabird, the red-footed booby, in a dynamic marine environment. *Marine Ecology Progress Series* **288**: 251–61.
17. Pinaud D. and Weimerskirch, H. 2005. Scale-dependent habitat use in a long-ranging central place predator. *Journal of Animal Ecology* **74**: 852–63.
18. Weimerskirch, H., Doncaster C.P. and Cuenotchaillet, F. 1994. Pelagic seabirds and the marine environment - foraging patterns of Wandering Albatrosses in relation to prey availability and distribution. *Proceedings of the Royal Society of London, B* **255**: 91–7.
19. Fayet, A.L., Freeman, R., Shoji, A., Padget, O. *et al.* 2015. Lower foraging efficiency in immatures drives spatial segregation with breeding adults in a long-lived pelagic seabird. *Animal Behaviour* **110**: 79–89.
20. Wade, H.M., Masden, E.A., Jackson, A.C., Thaxter, C.B. *et al.* 2014. Great Skua (*Stercorarius skua*) movements at sea in relation to marine renewable energy developments. *Marine Environmental Research* **101**: 69–80.
21. Young, H.S., Shaffer, S.A., McCauley, D.J., Foley, D.G. *et al.* 2010. Resource partitioning by species but not sex in sympatric boobies in the central Pacific Ocean. *Marine Ecology Progress Series* **403**: 291–301.
22. Orben, R.A., Irons, D.B., Paredes, R., Roby, D.D. *et al.* 2015. North or south? Niche separation of endemic Red-legged Kittiwakes and sympatric Black-legged Kittiwakes during their non-breeding migrations. *Journal of Biogeography* **42**: 401–12.
23. Navarro, J., Cardador, L., Brown, R. and Phillips, R.A. 2015. Spatial distribution and ecological niches of non-breeding planktivorous petrels. *Scientific Reports* **5**: 12164. Doi: 10.1038/srep12164.
24. Quillfeldt, P., Cherel, Y., Delord, K. and Weimerkirch, H. 2015. Cool, cold or colder? Spatial segregation of prions and Blue Petrels is explained by

differences in preferred sea surface temperatures. *Biology Letters* **11**: 20141090. Doi: 10.1098/rsbl.2014.1090.

## Chapter 9. How Seabirds Catch Food

1. Weimerskirch, H., Bishop, C., Jeanniard-du-Dot, T., Prudor, A. *et al.* 2016. Frigate birds track atmospheric conditions over months-long transoceanic flights. *Science* **353**: 74–8.
2. Weimerskirch, H., Le Corre, M., Ropert-Coudert, Y., Kato, A. *et al.* 2005. The three-dimensional flight of Red-footed Boobies: adaptations to foraging in a tropical environment? *Proceedings of the Royal Society of London, B* **272**: 53–61.
3. Weimerskirch, H. and Wilson, R.P. 1992. When do Wandering Albatrosses *Diomedea exulans* forage? *Marine Ecology Progress Series* **86**: 297–300.
4. Weimerskirch, H., Wilson, R.P. and Lys, P. 1998. Activity pattern of foraging in the Wandering Albatross: A marine predator with two modes of prey searching. *Marine Ecology Progress Series* **151**: 245–54.
5. Weimerskirch, H., Gault, A. and Cherel, Y. 2005. Prey distribution and patchiness: Factors in foraging success and efficiency of Wandering Albatrosses. *Ecology* **86**: 2611–22.
6. http://wildlifecomputers.com/our-tags/daily-diary/ (accessed 14 June 2017).
7. Grémillet, D., Prudor, A., le Maho, Y. and Weimerskirch, H. 2012. Vultures of the seas: Hyperacidic stomachs in Wandering Albatrosses as an adaptation to dispersed food resources, including fishery wastes. *PLoS One* **7**: e37834. Doi: 10.1371/journal.pone.0037834.
8. Dias, M.P., Alho, M., Granadeiro, J.P. and Catry, P. 2015. Wanderer of the deepest seas: Migratory behaviour and distribution of the highly pelagic Bulwer's petrel. *Journal of Ornithology* **156**: 955–62.
9. Dias, M.P., Romero, J., Granadeiro, J.P., Catry, T. *et al.* 2016. Distribution and at-sea activity of a nocturnal seabird, the Bulwer's Petrel *Bulweria bulwerii*, during the incubation period. *Deep-sea research. Part I, Oceanographic Research Papers* **113**: 49–56.
10. Pinet, P., Jaeger, A., Cordier, E., Potin, G. *et al.* 2011. Celestial moderation of tropical seabird behavior. *PLoS One* **6**: e27663. Doi: 10.1371/journal.pone.0027663.
11. Lewis, S., Benvenuti, S., Dall-Antonia, L., Griffiths, R. *et al.* 2002. Sex-specific foraging behaviour in a monomorphic seabird. *Proceedings of the Royal Society of London, B* **269**: 1687–93.

12. http://newsfeed.time.com/2012/08/01/watch-this-giant-bird-dives-150-feet-underwater-for-food/ (accessed 14 June 2017).
13. Tremblay, Y., Cook, T.R. and Cherel, Y. 2005. Time budget and diving behaviour of chick-rearing Crozet shags. *Canadian Journal of Zoology* **83**: 971–82.
14. Wilson, R.P., Vargas, F.H., Steinfurth, A., Riordan, P. *et al.* 2008. What grounds some birds for life? Movement and diving in the sexually dimorphic Galapagos Cormorant. *Ecological Monographs* **78**: 633–52.
15. Navarro, J., Votier, S.C. and Phillips, R.A. 2014. Diving capabilities of diving petrels. *Polar Biology* **37**: 897–901.
16. Shoji, A., Dean, B., Kirk, H., Freeman, R. *et al.* 2016. The diving behaviour of the Manx Shearwater *Puffinus puffinus*. *Ibis* **158**: 598–606.
17. See supplementary material listed by Navarro, J., Votier, S.C. and Phillips, R.A. 2014. Diving capabilities of diving petrels. *Polar Biology* **37**: 897–901.
18. Regular, P.M., Hedd, A. and Montevecchi, W.A. 2011. Fishing in the dark: A pursuit-diving seabird modifies foraging behaviour in response to nocturnal light levels. *PLoS One* **6**: e26763. Doi: 10.1371/journal.pone.0026763.
19. Wienecke, B., Robertson, G., Kirkwood, R. and Lawton, K. 2007. Extreme dives by free-ranging Emperor Penguins. *Polar Biology* **30**: 133–42.
20. Kokubun, N., Kim, J.H., Shin, H.C., Naito, Y. *et al.* 2011. Penguin head movement detected using small accelerometers: A proxy of prey encounter rate. *Journal of Experimental Biology* **214**: 3760–7.
21. Handley, J.M. and Pistorius, P. 2015. Kleptoparasitism in foraging Gentoo Penguins *Pygoscelis papua*. *Polar Biology* **39**: 391–5. Doi: 10.1007/s00300-015-1772-2. See also https://www.youtube.com/watch?v=i40i1VGANyM (accessed 14 June 2017).
22. Crook, K.A. and Davoren, G.K. 2014. Underwater behaviour of common murres foraging on capelin: Influences of prey density and antipredator behaviour. *Marine Ecology Progress Series* **501**: 279–90.
23. Regular, P.M., Hedd, A. and Montevecchi, W.A. 2011. Fishing in the dark: A pursuit-diving seabird modifies foraging behaviour in response to nocturnal light levels. *PLoS One* **6**: e26763. Doi: 10.1371/journal.pone.0026763.
24. Pütz, K., Wilson, R.P., Charrassin, J.-B., Raclot, T. *et al.* 1998. Foraging strategy of King Penguins (*Aptenodytes patagonicus*) during summer at the Crozet Islands. *Ecology* **79**: 1905–21.
25. Meir, J.U., Stockard, T.K., Williams, C.L., Ponganis, K.V. *et al.* 2008. Heart rate regulation and extreme bradycardia in diving Emperor Penguins. *Journal of Experimental Biology* **211**: 1169–79.

26. Grémillet, D., Wilson, R.P., Wanless, S. and Peters, G. 1999. A tropical bird in the Arctic (the cormorant paradox). *Marine Ecology Progress Series* **188**: 305–9.
27. Grémillet, D., Chauvin, C., Wilson, R.P., Le Maho, Y. *et al.* 2005. Unusual feather structure allows partial plumage wettability in diving Great Cormorants *Phalacrocorax carbo*. *Journal of Avian Biology* **36**: 57–63.
28. Quintana, F., Wilson, R.P. and Yorio, P. 2007. Dive depth and plumage air in wettable birds: The extraordinary case of the Imperial Cormorant. *Marine Ecology Progress Series* **334**: 299–310.
29. Wilson, R.P., Shepard, E.L.C., Gomez Laich, A., Frere, E. *et al.* 2010. Pedalling downhill and freewheeling up; a penguin perspective on foraging. *Aquatic Biology* **8**: 193–202.
30. Ponganis, P.J., Van Dam, R.P., Marshall, G., Knower, T. *et al.* 2000. Sub-ice foraging behavior of Emperor Penguins. *Journal of Experimental Biology* **203**: 3275–8
31. Sato, K., Ponganis, P.J., Habara, Y. and Naito, Y. 2005. Emperor Penguins adjust swim speed according to the above-water height of ice holes through which they exit. *Journal of Experimental Biology* **208**: 2549–54.
32. Chimienti, M., Cornulier, T., Owen, E., Bolton, M. *et al.* 2016. The use of an unsupervised learning approach for characterizing latent behaviors in accelerometer data. *Ecology and Evolution* **6**: 727–41.


## Chapter 10. The Clash: Seabird Interactions with People – Past, Present and Future


1. http://www.youtube.com/watch?v=-SDVV4Vz2kI (accessed 14 June 2017). *St. Kilda, Its People and Birds* (1908)
2. Dilley, B.J., Schoombie, S., Schoombie, J. and Ryan, P.G. 2016. 'Scalping' of albatross fledglings by introduced mice spreads rapidly at Marion Island. *Antarctic Science* **28**: 73–80.
3. Rodríguez, A., Rodríguez, B. and Negro. J.J. 2015. GPS tracking for mapping seabird mortality induced by light pollution. *Scientific Reports* **5**: 10670. Doi: 10.1038/srep10670.
4. https://vimeo.com/106024314 (accessed 14 June 2017).
5. http://www.hulldailymail.co.uk/jet-skier-appeals-riding-ban-disturbing-seabirds/story-26822641-detail/story.html (accessed 14 June 2017).
6. Auman, H.J., Ludwig, J.L., Giesy, J.P. and Colborn, T. 1998. Plastic ingestion by Laysan Albatross chicks on Sand Island, Midway Atoll, in 1994 and

1995. Pp. 239–44 in *Albatross Biology and Conservation* (G. Robertson and R. Gales (eds.)). Chipping Norton: Surrey Beatty & Sons.
7. See, for example, Chris Jordan's film *Midway: Message from the Gyre* at https://vimeo.com/25563376 (accessed 14 June 2017).
8. https://www.poetryfoundation.org/poems-and-poets/poems/detail/43997 (accessed 28 June 2017).
9. Thaxter, C.B., Ross-Smith, V.H., Bouten, W., Clark, N.A. *et al.* 2015. Seabird-wind farm interactions during the breeding season vary within and between years: A case study of Lesser Black-backed Gull *Larus fuscus* in the UK. *Biological Conservation* **186**: 347–58.
10. Desholm, M. and Kahlert, J. 2005. Avian collision risk at an offshore wind farm. *Biology Letters* **1**: 296–8. Doi: 10.1098/rsbl.2005.0336.
11. Ross-Smith, V.H., Thaxter, C.B., Masden, E.A., Shamoun-Baranes, J. *et al.* 2016. Modelling flight heights of Lesser Black-backed Gulls and Great Skuas from GPS: a Bayesian approach. *Journal of Applied Ecology* **53**: 1676–85.
12. Cleasby, I.R., Wakefield, E.D., Bearhop, S., Bodey, T.W. *et al.* 2015. Three-dimensional tracking of a wide-ranging marine predator: Flight heights and vulnerability to offshore wind farms. *Journal of Applied Ecology* **52**: 1474–82.
13. Collet, J., Patrick, S.C. and Weimerskirch, H. 2015. Albatrosses redirect flight towards vessels at the limit of their visual range. *Marine Ecology Progress Series* **526**: 199–205.
14. Bodey, T.W., Jessopp, M.J., Votier, S.C., Gerritsen, H.D. *et al.* 2014. Seabird movement reveals the ecological footprint of fishing vessels. *Current Biology* **24**: R514-R515. Doi: 10.1016/j.cub.2014.04.041.
15. Rollinson, D.P., Dilley, B.J. and Ryan, P.G. 2014. Diving behaviour of White-chinned Petrels and its relevance for mitigating longline bycatch. *Polar Biology* **37**: 1301–8.
16. Thiebot, J.-B., Demay, J., Marteau, C. and Weimerskirch, H. 2015. The rime of the modern mariner: Evidence for capture of yellow-nosed albatross from Amsterdam Island in Indian Ocean longline fisheries. *Polar Biology* **38**: 1297–1300.
17. Ratcliffe, N., Hill, S.L., Staniland, I.J., Brown, R. *et al.* 2015. Do krill fisheries compete with Macaroni Penguins? Spatial overlap in prey consumption and catches during winter. *Diversity and Distributions* 21: 1339–48.
18. Delord, K., Cotté, C., Péron, C., Marteau, C. *et al.* 2010. At-sea distribution and diet of an endangered top predator: Relationship between White-chinned Petrels and commercial longline fisheries. *Endangered Species Research* **13**: 1–16.

19. Tancell, C., Sutherland, W.J. and Phillips, R.A. 2016. Marine spatial planning for the conservation of albatrosses and large petrels breeding at South Georgia. *Biological Conservation* **198**: 165–76.
20. Lascelles, B.G., Taylor, P.R., Miller, M.G.R., Dias, M.P. *et al.* 2016. Applying global criteria to tracking data to define important areas for marine conservation. *Diversity and Distributions* **22**: 422–31.
21. Péron, C., Grémillet, D., Prudor, A., Pettex, E. *et al.* 2013. Importance of coastal Marine Protected Areas for the conservation of pelagic seabirds: The case of Vulnerable Yelkouan Shearwaters in the Mediterranean Sea. *Biological Conservation* **168**: 210–21.
22. Meier, R.E., Wynn, R.B., Votier, S.C., McMinn Grivé, M. *et al.* 2015. Consistent foraging areas and commuting corridors of the critically endangered Balearic Shearwater *Puffinus mauretanicus* in the northwestern Mediterranean. *Biological Conservation* **190**: 87–97.
23. https://www.cbd.int/sp/targets/rationale/target-11 (accessed 14 June 2017). See also United Nations Framework Convention on Climate Change Decision VII/15.
24. Regular, P., Montevecchi, W., Hedd, A., Robertson, G. *et al.* 2013. Canadian fishery closures provide a large-scale test of the impact of gillnet bycatch on seabird populations. *Biology Letters* **9**: 20130088. Doi: 10.1098/rsbl.2013.0088.
25. Daunt, F., Wanless, S., Greenstreet, S.P.R., Jensen, H. *et al.* 2008. The impact of the sandeel fishery closure on seabird food consumption, distribution, and productivity in the northwestern North Sea. *Canadian Journal of Fisheries and Aquatic Sciences* **65**: 362–81.
26. https://www.rspb.org.uk/Images/RSPB_Scotland_SPA_report_May2014_tcm9-369474.pdf (accessed 14 June 2017).
27. www.theguardian.com/environment/2016/sep/23/terns-migration-alaska (accessed 14 June 2017).
28. https://www.youtube.com/watch?v=_tdO72TFXz0 (accessed 14 June 2017). Puffin chick eating a butterfish *Poronotus triacanthus*.
29. Jenouvrier, S., Caswell, H., Barbraud, C., Holland, M. *et al.* 2009. Demographic models and IPCC climate projections predict the decline of an emperor penguin population. *Proceedings of the National Academy of Sciences USA* **106**: 1844–47.
30. Huntley, B., Green, R.E., Collingham, Y.C. and Willis, S.G. 2007. *A climatic atlas of European breeding birds*. Barcelona: Lynx Edicions.

31. Hodgson, J.C., Baylis, S.M., Mott, R., Herrod, A. *et al.* 2016. Precision wildlife monitoring using unmanned aerial vehicles. *Scientific Reports* **6**: 22574. Doi: 10.1038/srep22574.
32. Fretwell, P.T., LaRue, M.A., Morin, P., Kooyman, G.L. *et al.* 2012. An Emperor Penguin population estimate: The first global, synoptic survey of a species from space. *PLoS One* **7**: e33751. Doi: 10.1371/journal.pone.0033751.

# INDEX

# Seabird Locations

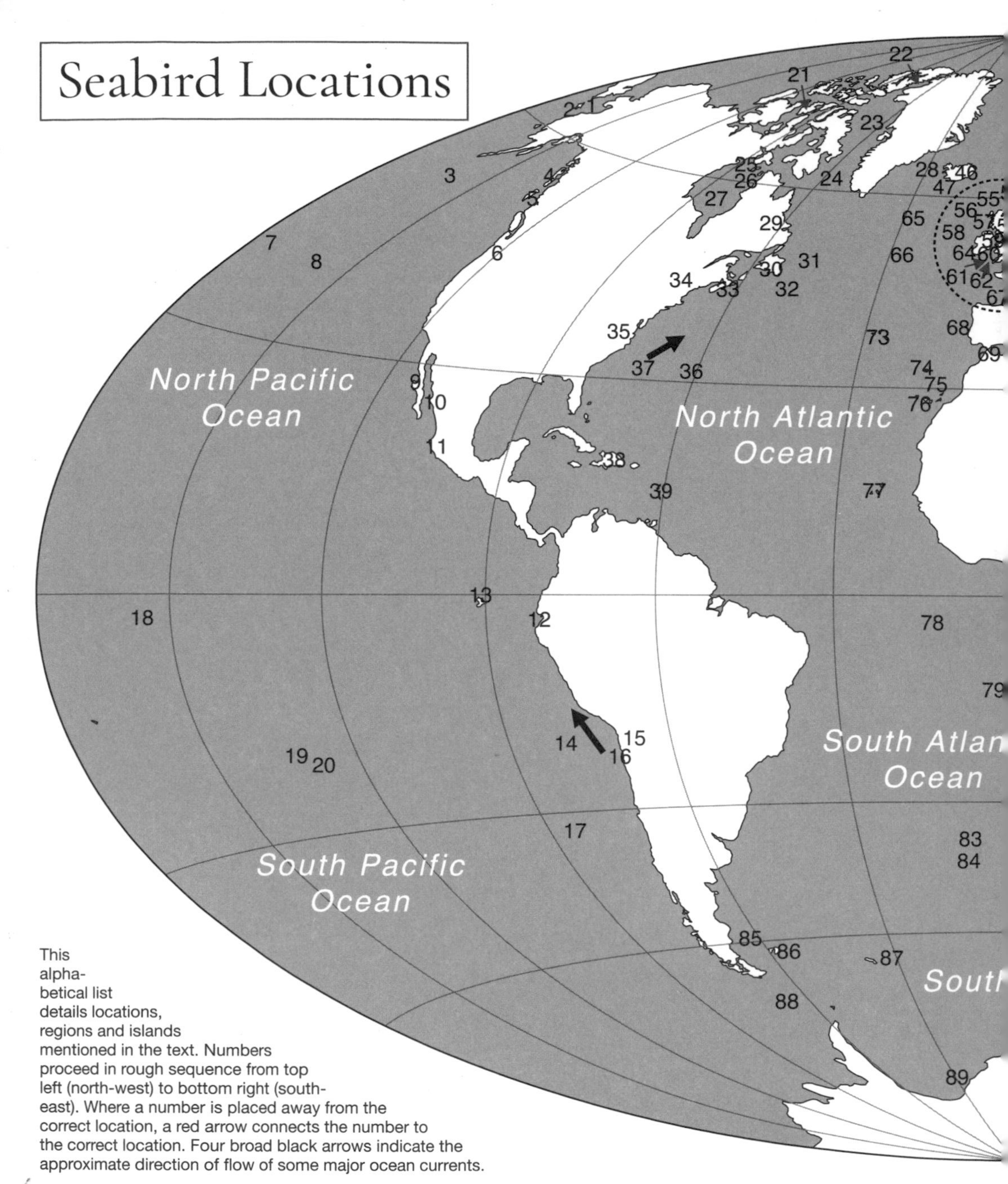


This alphabetical list details locations, regions and islands mentioned in the text. Numbers proceed in rough sequence from top left (north-west) to bottom right (south-east). Where a number is placed away from the correct location, a red arrow connects the number to the correct location. Four broad black arrows indicate the approximate direction of flow of some major ocean currents.

Acapulco 11
Agulhas Current 95
Agulhas slope 96
Ailsa Craig 58
Albatross Island 107
Alejandro Selkirk, Juan Fernández archipelago 17
Aleutian Islands 3
Amsterdam Island 93
Antofagasta 16
Arabian Sea 99
Ascension Island 78
Atacama Desert 15
Azores 73
Baffin Bay 23
Baja California 9
Barents Sea 44
Bass Rock, Firth of Forth 53
Bay of Bengal 101
Bay of Biscay 67
Bempton Cliffs, Yorkshire 54
Benguela Current 80
Bering Sea 1
Berlengas 68
Bermuda 36
Bjørnøya (Bear Island) 43
Bugio, Desertas 74
Canary Islands 76
Cape of Good Hope 82
Cape Verdes Islands 77
Chagos 100
Charlie-Gibbs Fracture Zone 66
Christmas Island 102
Coats Island 25
Codfish Island 115
Copeland Island 59
Corsica 71
Dassen Island 81
Davis Strait 24
Denmark Strait 28
Digges Island 26
Dogger Bank 52
Dominica 39
Drake Passage 88
Ecuador 12
Ellesmere Island 22
Europa 97
Fair Isle 50
Falkland Islands 86
Farne Islands 54
Fiji 121
Forrester Island 4
Franz Josef Land 40
Galápagos Islands 13
Gold Coast 105
Gough Island 84
Grand Banks 32
Grassholm 60
Great Australian Bight 110
Great Barrier Island 119
Great Barrier Reef 104
Gulf of Mexico 10
Gulf Stream 37
Haida Gwaii 5
Henderson Island 19
Hispaniola 38
Hokkaido 124
Hudson Bay 27
Humboldt Current 14
Iceland 46